AF879227

Advanced Gasification

Solar Energy R&D in the European Community

Series E:

Energy from Biomass

Volume 8

Solar Energy R&D
in the European Community

Series E Volume 8
Energy from Biomass

Advanced Gasification

Methanol Production from Wood - Results of the EEC Pilot Programme

edited by

A. A. C. M. BEENACKERS
University of Groningen, Groningen, The Netherlands

and

W. VAN SWAAY
Twente University of Technology, Enschede, The Netherlands

D. Reidel Publishing Company
A MEMBER OF THE KLUWER ACADEMIC PUBLISHERS GROUP
Dordrecht / Boston / Lancaster / Tokyo

for the Commission of the European Communities

Library of Congress Cataloging in Publication Data
Main entry under title:

Advanced gasification.

(Solar energy R & D in the European community.
Series E, Energy from biomass; v. 8)
Includes index.
1. Gas-producers–Addresses, essays, lectures. 2. Wood as fuel–Addresses, essays, lectures. I. Beenackers, A. A. C. M. II. Van Swaaij, W. P. M. van. III. Series.
TP762.A38 1986 662'.8 86-483
ISBN 90-277-2212-9

CIP

Publication arrangements by
Commission of the European Communities
Directorate-General Information Market and Innovation, Luxembourg

EUR 10407
© 1986 ECSC, EEC, EAEC, Brussels and Luxembourg

LEGAL NOTICE
Neither the Commission of the European Communities nor any person acting on behalf of the Commission is responsible for the use which might be made of the following information.

Published by D. Reidel Publishing Company
P.O. Box 17, 3300 AA Dordrecht, Holland

Sold and distributed in the U.S.A. and Canada
by Kluwer Academic Publishers,
190 Old Derby Street, Hingham, MA 02043, U.S.A.

In all other countries, sold and distributed
by Kluwer Academic Publishers Group,
P.O. Box 322, 3300 AH Dordrecht, Holland

All Rights Reserved
No part of the material protected by this copyright notice may be reproduced or utilized in any form or by any means, electronic or mechanical, including photocopying, recording or by any information storage and retrieval system, without written permission from the copyright owner.

Printed in The Netherlands

PREFACE

Advanced gasification, the subject of this present book, was an important component in the European Community's R&D Programme on Energy from Biomass. The Commission had issued a particular call for tenders and had selected four contractors in France, Germany, Italy and the United Kingdom for support. (See also proceedings of the workshop on biomass pilot projects, published in Volume 2 of this same series E).

Later on, four complete full-sized pilot projects with a throughput of approximately 1 tonne of biomass per hour were built up and an energy-rich synthesis gas was produced. Some of the gas could be successfully converted into methanol.

It is important to note that all four projects proved eventually to be technically viable although at least two of them were based on completely new ideas and included high development risks. As a result, a set of new gasification technologies of a very high standard have become available in Europe. The main emphasis was on gasification of wood but nevertheless all plants are also of interest for the gasification of coal and municipal waste.

It should be emphasized that this pilot programme has become a good example of European cooperation of large industrial companies. In the frame of the Community's programme, the four contractors who included some of Europe's biggest companies for mechanical engineering accepted to cooperate effectively.

The pilot programme also included a few minor projects on the same subject which were exploratory. Here too interesting results were achieved which are presented in this book.

The book also includes a general overview of the programme which has been prepared by the Commission's experts in this area, Professors Van Swaaij and Beenackers.

This book should become a valuable reference for all those interested in advanced gasification technologies. I would like to thank all participants in this programme for their cooperation.

W. Palz
Head of the Biomass R&D Sub-programme

CONTENTS

REVIEW PAPER

Introduction to the biomass to synthesis gas pilot plant programme of the C.E.C. and a final evaluation of its results

A.A.C.M. Beenackers and W.P.M. van Swaaij

INTRODUCTION TO THE BIOMASS TO SYNTHESIS GAS PILOT PLANT PROGRAMME OF THE C.E.C. AND A FIRST EVALUATION OF ITS RESULTS

A.A.C.M. BEENACKERS* and W.P.M. VAN SWAAIJ**

Summary and Conclusions

Four pilot plant projects aiming at the production of synthesis gas suitable for methanol manufacturing were supported by the Commission of the European Communities during a three year programme lasting from January 1982 to the end of 1984. An introduction and an first evaluation of its results will be given in this chapter, which should be considered as the personal view of the authors only.
Two smaller somewhat related pilot plant projects presented in this volume are also shortly described.

The most important properties of the four pilot plants, which ranged in design capacity from 4.8 - 12 tons dry wood/day, are given in Table IV. For each of these projects a checklist has been worked out with the more important and/or critical development items. The results have been condensed in Table VII. From the most successful projects synthesis gas was actually converted to methanol in a methanol pilot unit of Lurgi.

The Creusot Loire plant was able to run for more than 24 hrs on the CEC-bonus conditions given in Table III. Positive points of the process are excellent product gas quality for methanol synthesis, high degree of conversion of biomass, high hydrocarbon conversion and good prospects for pressurized operation; a pressurized pilot plant of 60 ton dry wood/day is planned to be constructed at Clamecy. Uncertainties lay in the area of heat recovery from the product gas (if desired) and possibly removal of entrained ash while the high oxygen consumption is a disadvantage.

The Lurgi Circulating Bed pilot plant also operated satisfactorily for more than 24 hrs nearly at CEC-bonus conditions. Only the methane content was a bit too high due to the relatively low operation temperature. Positive points of the process are high specific unit capacity; large operation flexibility, a relatively simple single bed system with proven scaling-up abilities in other areas. Due to the lower than anticipated bed temperature the unit capacity was lower than foreseen and the methane content of the product gas a bit too high. Unfortunately no data at higher temperature have been published yet and stable operation at these conditions remains to be proven. Heat recovery from the product gas was not piloted in this project. A disadvantage is the use of oxygen which will still increase a bit if synthesis gas according to the CEC bonus conditions is to be produced.

* Groningen University, Laboratory for Chemical Engineering, Nijenborgh 16, 9747 AG Groningen, The Netherlands

** Twente University of Technology, Laboratory for Chemical Reaction Engineering and Industrial Processes, P.O. Box 217, 7500 AE Enschede, The Netherlands

The John Brown Wellman ODG pilot plant made a longest testrun of 9.5 hrs but the gas composition was still far away from the CEC bonus conditions due to the high methane content. It is not completely sure that the product gas was produced via the complex chemical mechanisms on which the process is based; possibly part of it was produced by other mechanisms like steam gasification. It is a pity that simple steam gasification was not tried in the test series. Positive points of the process are the use of only air as a gasification agent and the elegant solution of the solids circulation system. Disadvantages are the complex chemistry of the process, which is not yet completely proven, the high hydrocarbon content of the gas (in case methanol production is desired), the high specific stone consumption and the potential danger of sulphur emission. Not enough data are available yet to make accurate balances, specially because also a small amount of coal was co-fired in the plant during the testrun.

The Italenergie/AGIP pilot plant had only made short duration runs (2 hrs) when this report was written. During the development and construction several aspects became clear however. An advantage is the use of indirect heating by which oxygen can be avoided in the primary gasification step although oxygen will have to be used to bring the product gas to methanol synthesis gas specifications in the secondary gasifier. A second positive point is the testing of all important heat exchangers in the pilot plant. Potential problems are related to the high temperature heating wall around the gasifier, fouling problems in the primary gasifier system and recycle streams and only partial conversion of the char.

Two additional pilot projects with slightly different scope have been supported by the CEC.

The AVSA project (University of Brussels) concerns a double fluid bed steam/air gasifier in a special compact lay-out avoiding solids transport lines by circulating solids between gasifier and combustor by an ingeneous kind of manipulated "free convection". A large scale pilot plant (200 kg wood/hr) has been constructed and the combustor section has already been operated. Presently the gas exchange rate between gasifier and combustor is still much too high (20-40%) and this will require further development. The product gas will certainly contain tar and methane.

The Twente University of Technology project on hydrogen recovery aims at recovery and pressurizing hydrogen from low Joule gases obtained by simple air gasification in a continuous process using an hydridible metal slurry in an absorption/desorption cycle. If this system can be operated on a large scale both oxygen and complex double bed gasification can be avoided. Together with existing absorption processes for CO and/or shift processes both synthesis gas and pure hydrogen can be produced. Laboratory experiments look promising and a pilot plant will be operated in the near future.

Future developments improving the application prospects both for the gasification step and the methanol process are shortly discussed. The paper starts with a condensed overview of methanol synthesis processes as far as it is relevant to select an optimal up-stream gasification process.

1. INTRODUCTION

Biomass is a renewable, relatively clean, source of energy with a low sulphur content and with less environmental problems as e.g. coal. Main disadvantage probably is its scattered availability and moreover it is still a solid fuel with all the well known disadvantages of solids as fuels relative to gases and liquids. Therefore many types of conversion processes for wood to other energy carriers like liquids, gases and electricity have been developed.

Liquid fuels can be made by pyrolysis, direct liquefaction or via the so called indirect route where the wood is first converted to synthesis gas which then is converted to a liquid mixture of hydrocarbons or alcohols. Comparative feasibility studies have shown that the indirect liquefaction of wood to methanol presently probably is probably the most promising process for generating liquid fuels from biomass.

Industrial production of methanol from synthesis gas already stems from 1920 or earlier and has made a long development to a modern process. Production of synthesis gas from wood is also very old, probably more than 250 years [1]. Therefore, one could think that the production of methanol from wood would not require additional research and development work. It is not true, mainly due to the difficulty in producing efficiently and economically a clean synthesis gas, preferentially under pressure. There is still no such commercial process available although a few processes are in a stage of advanced development, most of them in the E.C. with financial support of the C.E.C. Solar Energy R&D Research Fund. A memorable event in this respect is the production by Lurgi, Frankfurt, of small amounts of methanol from wood through the indirect route, probably for the first time ever, at least in Europe, in June 1983. To provide some background information we start with a discussion on the various process routes and reactor types that could be used for producing methanol from biomass via synthesis gas. For a more detailed analysis on the possibilities, economics, and for a state of the art in the world the reader is referred to our previously published reviews (see e.g. [2], [3]). In this introduction the results of the current C.E.C. pilot projects on gasification of biomass for synthesis gas production will be analysed whilst unsolved problem areas will be identified.

We start, however, with a short summary on methanol production technology because this process dictates the optimal synthesis gas composition and pressure and therefore has a significant impact on the question what type of gasification process is optimal with respect to downstream methanol production.

1.1 Methanol production

A modern large scale methanol plant may have a capacity of a 1000 tons/day of methanol and is carefully energy integrated with the synthesis gas production plant.

Several processes are available but the processes of ICI and Lurgi are most commonly used [4,5]. Both operate at temperatures between 220 °C and 300 °C, use a copper based catalyst and operate between 50-100 bar. Figure 1 gives a schematic view of the reactor section. Because methanol production via

$$CO + 2H_2 \rightleftarrows CH_3OH \qquad - 90.77\ \text{kJ/mol} \qquad (1)$$

with usually some co-production via

$$CO_2 + 3H_2 \rightleftarrows CH_3OH + H_2O \qquad - 49.52\ \text{kJ/mol} \qquad (2)$$

is overall strongly exothermic, heat removal from the conversion/recycle loop and efficient utilization of this heat and of the bleedstream gas are crucial

factors in the process economics. The processes of ICI and Lurgi differ in a few details of which the heat removal from the reactor (by direct raising of steam (Lurgi) or by the cold shot technique (ICI)) is the most significant one [2].
One of the possible schemes for wood (or other biomass) based methanol production is given in a simplified form in Figure 2. It is based on an oxygen blown pressurized gasifier but of course also other types producing a suitable synthesis gas can be used.

After the gasification step some clean-up of the gas from dust and residual tar will be required. In the CO-shift reactor the ratio of CO/H_2 can be adjusted by steam addition via the catalytically enhanced slightly exothermic reaction

$$CO + H_2O \rightleftarrows CO_2 + H_2 \qquad -40.5 \text{ kJ/mol} \qquad (3)$$

Then, prior to the synthesis step, CO_2 and H_2S have to be removed. Some CO_2 can be tolerated, is in fact even essential to get the methanol synthesis going, but H_2S is a catalyst poison and has to be removed completely. Wood only contains very small amounts of sulphur (<50 ppm) and therefore the "sweetening step" (CO_2 and H_2S removal) can be relatively cheap compared to e.g. a methanol from coal process.

A major problem with biomass, at least in Europe, is its scattered availability. Usually the production areas will be too small to allow for a typical methanol plant capacity of a thousand tons per day.

The economy of scale of the whole methanol complex is relatively important [2]. This is mainly caused by the methanol synthesis plant. Therefore addition of the crude synthesis gas to the feedstock of a large scale (e.g. natural gas based) methanol complex could ease the large scale requirement for biomass gasification. On the other hand, the methanol plant itself could be improved and/or adapted to the wood gasifier or, generally, to small scale operation. Firstly, it should be realized that it may not be necessary to produce a chemical grade methanol if the product is to be used as a fuel and the purification step may be omitted or simplified. New methanol processes may produce more economically a fuel grade product; Indeed, such new processes have been announced recently by IFP (France) and Lurgi (FRG) [6].

Another new development of the last decade is the so called Chem Systems [7,8] LPMeOH process. Here, an inert liquid phase (typically an aliphatic C_{12}-C_{21} or an aromatic C_{10}-C_{11} fraction) is used to suspend and cool the catalyst. Two versions are under consideration: a liquid fluidized bed version with particles typically in the mm range and an entrained bed version with catalyst particle dimensions in the micron range. It is claimed that these processes have a higher flexibility towards fluctuations in the feedstock, better temperature control, a higher conversion per pass (reduced recycle costs) and the possibility to accept a feed with a wide variety of CO/H_2 ratios.

Factors like catalyst stability, life expectation of catalyst, attrition/erosion resistance and process economics, presently are investigated by Air Products under a DOE contract. The investigations are carried out in a 5 tons methanol per day pilot plant facility at Laporte (Texas, USA). Some fundamental chemical reaction engineering research on this process recently started at Groningen University, The Netherlands.

A typical synthesis gas suitable for methanol production is defined in Table I.

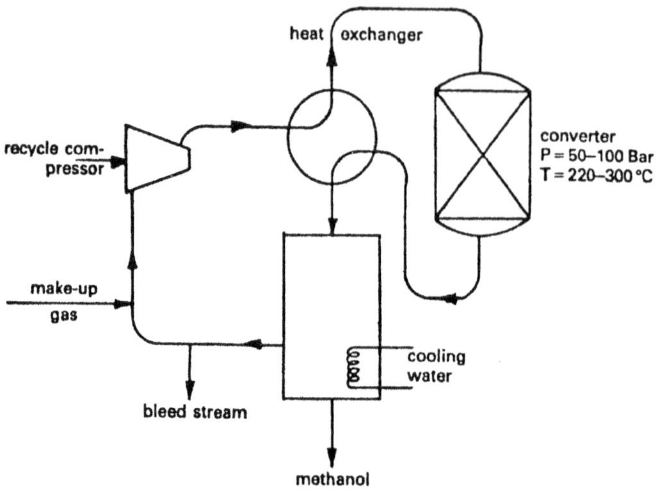

Figure 1 Methanol synthesis loop

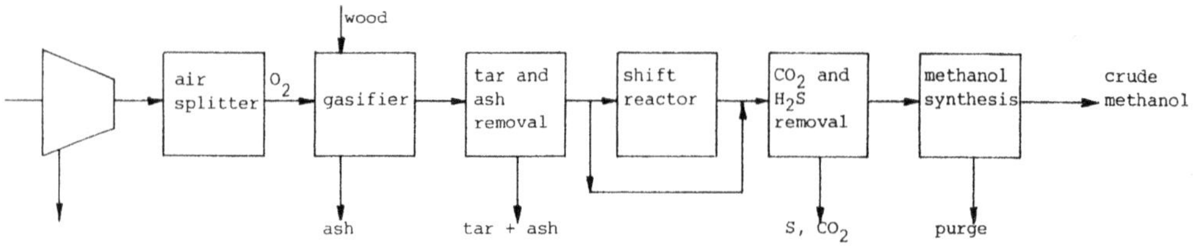

Figure 2 Simplified scheme of a possible methanol synthesis from wood

Table I. Typical synthesis gas required for methanol production; $H_2/(2CO + 3CO_2) \approx 1$

Component	Conc. [mol%]
$CO + H_2$	maximal (typically >70)
CO_2	2-10
$CH_4 + N_2$	minimal (typically < 3)
H_2S	$< 10^{-5}$

1.2 Reactor types and process conditions

To produce a suitable synthesis gas from wood several different routes can be chosen. None of these are completely proven on a commercial scale, but some have reached the demonstration scale of 10-50 tons of wood gasification per day.

Ideally, only useful synthesis gas components should be produced, e.g. by:

$$C_6H_{10}O_5 + \frac{1}{2} O_2 \rightarrow 6CO + 5H_2 \qquad (4)$$

Apart from the fact that this reaction is endothermic it can not be realized with 100% selectivity. Already during drying and heating-up of the biomass a broad variety of pyrolysis products is produced such as methane and higher hydrocarbons, water and carbon dioxide. Furthermore, additional water and carbon dioxide will generally be produced as by-products from undesirable combustion reactions proceeding simultaneously with the gasification process while some additional methane can be formed by hydrogenation. Therefore, the composition of the raw synthesis gas will differ substantially from what eqn. (4) may suggest. Moreover, the actual composition will not only depend on the choice of gasifying agent (steam or oxygen), but also on both the reactor type selected and the operating conditions for the gasifier. We, therefore, can distinguish two main factors by which the various gasification processes can be characterized, i.e.:

- process conditions
- reactor type

This is done in form of a matrix in Table II. Both the particularly promising combinations and the CEC supported project areas are indicated in this table. For a detailed discussion on the merits and disadvantages of the various process conditions and reactor types, the reader is referred to one of our previous reviews on this topic [2,3].

2. THE PILOT PLANT PROGRAMME SUPPORTED BY THE C.E.C.

Early 1981, the Commission of the European Communities published a call for tenders on the development of pilot plants for the production of synthesis gas from wood at a capacity of 10-20 tons wood/day aiming at methanol production [9]. Continuous downstream production of methanol from the synthesis gas produced was not part of the pilot plant program as this technology was considered to be proven already.

A special financial bonus was announced to be available for those contractors that could produce, during a controlled 24 hrs testrun, a synthesis gas of a specified quality suitable for methanol synthesis. These bonus conditions are given in Table III.

process conditions \ reactor type	co-current packed bed	counter-current packed bed	cross current packed bed	fluid bed	circulating fast bed	powder flame reactor	double fluid bed	exotic reactor types
1. Atmospheric air	±	±	-	±	±	±		±
2. Atmospheric steam (+air)(electricity)	±	-	-		-	-		-
3. Atmospheric oxygen/ steam							-	
4. Pressurized air	-	-	-	-	-	-	±	-
5. Pressurized steam (+air)	-	-	-	-	-	-	±	-
6. Pressurized oxygen/ steam					±		-	±
	A	B	C	D	E	F	G	H

TABLE II Reactor types and process conditions for gasification of wood to synthesis gas

- \- no interest
- ± little interest and development
- \\\\\\ important research/dev. areas
- /// EC supported projects

Table III. Bonus conditions in the CEC pilot plant programme

Duration of the testrun	$\geq$ 24 hrs
Capacity, wood 20% moisture	$\geq$ 400 kg/hr
dry wood	$\geq$ 320 kg/hr
(H_2 + CO), volumetric* percentage	$\geq$ 70%
Methane + HC + inert*	$\leq$ 3%

*After correction for purge flows for instrument protection

Nine proposals were submitted to the Commission and discussed during the EC Workshop of 22 October 1981 in Brussels [10]. Shortly thereafter four proposals were selected on basis of optimal development potential. It meant the actual start of a three year development program of which the final reports are presented in this book.

The contractors involved in the programme, the process principles and the pilot plant characteristics are all summarized in Table IV.

2.1.1 Reactor types chosen

As can be seen from Table IV all pilot plants are based on one or two fluid bed reactors thus providing the feedstock flexibility required to process wood chips or other biomass fragments which are generally difficult to handle in other type of reactors, specially on a large scale of operation. Moving bed reactors would require severe feedstock preselection and/or complex and costly feedstock preparation steps. Powder flame reactors would require a costly milling step as discussed elsewhere [2]. Nevertheless, the secondary gasifier of the Creusot Loire plant and to a certain extent the Lurgi Circulating Bed plant have incorporated powder flame reactor characteristics.

2.1.2 Gasifying agent

The processes of both Lurgi and Creusot Loire use mixtures of steam and oxygen to gasify the wood. Agip/Italenergie basically uses steam and air but still needs oxygen in the secondary gasifier. Due to the special features of the John Brown/Wellman process, only air (and possibly a small amount of steam) reportedly will be required for this process. It should be realized, however, that for the latter process it has not been demonstrated yet that complete gasification to synthesis gas can indeed be effected without using any oxygen because, at the present state of development, the product gas still contains substantial amounts of methane and other pyrolysis products. Therefore, major process improvements still have to be realized before a secondary (oxygen consuming) gasification or a steam reforming step safely can be skipped from a technically feasible design.

2.1.3 Scale of operation

The proven scale of operation (see table IV) is different for each of the four pilot projects but is always large enough to get all main technical problems at least identified. Nevertheless, none of them is probably large enough to allow for a safe extrapolation to the commercial scale unless additional experience from a similar reactor in a related application is available. For the Creusot Loire process a much larger pilot plant is planned. In this pilot plant also pressurized operation will be tested (± 15 bar).

Consortium or Company	pres-sure [bar]	design capacity [ton dry wood/h]	reactor type	gasification agent	duration testrun [hrs]	syngas methane + HC content [vol %]	proven actual capacity [ton dry wood/hr]	next step	oxygen and steam consumption [kg/kg dry wood]
Creusot Loire	1	0.320	fluid bed plus secondary empty tube gasifier	oxygen/steam for primary and secondary gasifier	24	0.6	0.350	2.5 t/h 15 bar Clamecy	O_2 : 0.57 H_2O: 0.08
Lurgi	1	0.320	circulating (fluid) bed gasifier	oxygen/steam	36	6.5	0.191	not yet announced	O_2 : 0.453 H_2O: 0.02
John Brown/Wellman	1	0.44	double fluid bed with chemically active solids	air	9.5	12.5	± 0.4 + 0.04 ton/h coal	not yet announced	only air
Italenergie/AGIP	1	0.5	primary fluid bed steam gasifier heated through wall plus secondary gasifier	steam for primary gasifier oxygen for secondary gasifier	0	*	0.25	not yet announced	only air **

TABLE IV Main characteristics of the four CEC supported gasification pilot plant projects.

* no accurate values available yet.

** secundary gasifier not yet in operation.

2.1.4 Criteria for success of the pilot projects

To put the results of the pilot plant programme into perspective we will discuss for each project how the difficulties of wood gasification have been solved and to what extent the solutions were successful. The projects can be considered to be a complete success if all technical aspects were found to be satisfactory and an economical evaluation shows favourable prospects for the full scale operation. Unfortunately, no pre-feasibility studies are available yet to allow for a quantitative comparison of the relative merits of the four different processes if they would be operated at a full scale. Moreover, the number of variables involved and the uncertainties inherent in comparing unproven technologies will prohibit accurate comparison of wood gasification plants coupled with gas treating units and methanol synthesis plants with, as a rule, different degrees of possible integration. Elsewhere [2], we concluded that such comparative studies may predict a spread in production costs from the various methanol to wood processes of no more than 20% which is too small to be of significance in the light of the still existing uncertainties in both actual capital and operation costs. Nevertheless, if the use of oxygen can be eliminated, there may be a definite decrease in the production costs of the synthesis gas and probably also in the costs of the methanol produced [2]. Therefore, though with some hesitation, this economical factor can be translated into a technical criterium.

A successful testrun of 24 hrs is a direct and relatively reliable indication that at least the serious short term technical problems have been solved. Therefore, this criterium has been included in the CEC bonus conditions. On the other hand, typical long term problem areas such as erosion/corrosion, fauling etc. will not be detected in a run of only 24 hrs. Moreover, not all items of an integrated full scale plant have been piloted in the projects (e.g. heat exchangers). We made a short list of 10 criteria for checking the development of a synthesis gas from wood process (table V). Before we check the different items of this list for each of the pilot projects we will first discuss the criteria seperately.

TABLE V 10 criteria for checking the development of a synthesis gas from wood process

a. Introduction of feedstock in the bed
b. Danger of ash melting/agglomeration of ash
c. Incomplete gasification due to a low temperature and/or a short residence time
d. Methane and other pyrolysis products in product gas
e. Entrainment of ash and char particles
f. Use of expensive oxygen
g. Safe and effective separation between gasifier and combustor (for double fluid beds only)
h. Heat transfer between gasifier and combustor
i. Recovery of sensible heat from product gas
j. Prospects for pressurization

a. Introduction of the feedstock in the fluid bed

Although this may seem to be a relatively simple matter, it is in fact a difficult development item. Due to its fibrous nature wood may easily block feeding devices. Moreover, because the feeding device is connected to a hot reactor, backmixing of heat into the feeder may cause partial pyrolysis of wood already in the feeder which may greatly promote the blocking. To provide gas-tightness and favourable solids flow often two or three separate devices in series are required.

b. **Danger of ash melting/agglomeration**

Wood ash may melt or agglomerate at the bed temperature or in the hot spots in the fluid bed (e.g. near the oxygen inlets). In some cases it may form a low melting eutecticum with the outer layers of inert fluidization material within the fluid bed. Agglomeration may be eminent only after several hours of operation and because the ash components may vary with the type of feedstock it remains a difficult item probably for all fluid beds operating at a temperature higher than 800-900 °C.

c. **Incomplete gasification due to a low temperature and/or a short residence time**

To alleviate problems with materials of construction and with ash melting and/or agglomeration the designer is often tempted to select a low reactor temperature. However, if the gasification temperature is low then also the char gasification rate will be low and this problem at best can only be partially counteracted by char recycling. A second problem occurring at low temperatures is the low conversion rate of the hydrocarbons formed in the critical wood pyrolysis step.

d. **Methane and other pyrolysis products in the product gas**

In pyrolysis of wood unavoidedly hydrocarbons are produced, specially methane. The higher the reactor temperature the less methane and other hydrocarbons will be formed and moreover the rate of conversion of these products to CO and H_2 by secondary gasification or reforming reactions is faster. Table VI gives the gas composition of both the Creusot Loire and the Lurgi pilot plant. As can be seen from this table there is a large difference in methane content which is due to the difference in the highest temperature in the process. The methane and other hydrocarbons conversion rate cannot be calculated beforehand because of unknown reaction kinetics. Therefore the actual methane and hydrocarbon concentrations can only be obtained experimentally. Increasing temperature and residence time will, however, improve the conversion of these components whilst, with decreasing temperatures below 800 °C substantial amounts of CH_4 will be obtained irrespective of the residence time because of thermodynamic equilibrium considerations. The other gaseous components CO, H_2, CO_2 and H_2 use to be close to the chemical equilibrium that can be calculated from the Schläpfer model (table VI). This model is based on mass and heat balances in combination with the water shift reaction equilibrium calculated at the reactor outlet temperature:

$$CO + H_2O \rightleftarrows CO_2 + H_2$$

Gas compositions thus calculated by the authors are included in table VI. Indeed, these theoretical data are very close to the composition observed experimentally.

TABLE VI Typical dry gas compositions for the Creusot Loire and Lurgi gasification pilot plants compared to those calculated from the modified Schläpfer model.

	Creusot Loire		Lurgi	
% vol	experimental T ≂ 1300 °C	calculated* T = 1300 °C	experimental T ≂ 750 °C	calculated* T = 750 °C
H_2	30.8	31.0	33.7	33.0
CO	39.3	41.2	30.0	31.3
CO_2	25.0	23.3	26.6	26.0
CH_4	0.6	0.6	4.9	4.9
C_2 etc.	-		1.6	1.6
N_2 + A	4.3	3.9	3.2	3.2
Heatloss		9%		5%

* Calculated by the authors from mass and heat balances assuming watershift reaction at equilibrium and assuming CH_4 + HC + N_2 as in experiments Assumed to be equal to experimental data

e. **Entrainment of ash and char particles**

Usually a substantial fraction of the char produced from the wood in the gasifier will be entrained with the fluidizing gas together with the ash. In most cases these solids will have to be recovered from the gas and at least partially be recycled. Another solution may be secondary gasification. Ash always will have to be recovered somewhere in the system and this ash should preferably contain minor amounts of char only.

f. **Use of expensive oxygen**

As we discussed elsewhere [2] the use of oxygen contributes largely to the operating costs of a gasification process. Therefore, processes in which the use of oxygen is avoided certainly have an important advantage which then still has to be judged against any disadvantage possibly caused by the fact that no-oxygen processes usually are more complex both in construction and operation.

g. **Safe and effective separation between gasifier and combustor (for double fluid beds only)**

If the solids present in a double fluid bed system are used as a heat carrier, technical solutions must be found to allow for a sufficiently fast circulation between the gasifier and the oxidiser (or combustor/regenerator) whilst maintaining an effective gas barrier between these two fluid beds. If, alternatively, the gasifier is heated up indirectly, then the problems are in finding the appropriate materials and geometry allowing for a sufficient and often considerable amount of heat to be transferred at high temperatures on both sides of the heat exchanging area, all in the hostile environment of a fluid bed. Failure of the wall may result in a dangerous situation.

h. **Heat transfer between gasifier and combustor (for double fluid beds only)**

For the different types of double fluid bed gasifiers both the heat transfer rate and its control are critical design and operation parameters.

i. **Recovery of sensible heat from the product gas**

If heat exchangers are planned in the full scale operation to recover heat of a dust and/or tar loaded product gas, this cannot be considered to be a proven technology but should be piloted.

j. **Prospects for pressurization**

Pressurization of the gasification process may reduce overall wood to methanol production costs [2]. For some processes pressurization will be possible for others difficult or impossible. Specially if delicate pressure balances have to be maintained between fluid bed sections pressurization will be a difficult matter. The different pilot projects will now be discussed separately using the checklist as a guideline. Table VII gives an overview of the results of the criteria applied to the four pilot plants.

2.2 Creusot Loire

Short general description [11]

This process consists of a two stage gasifier with a steam/oxygen mixture as gasification agent (Fig. 3). The first stage is a more or less classical fluidized bed gasifier. Due to the relatively low gasification temperature, the primary gas still contains a lot of pyrolysis products (tar, methane, etc.) together with unconverted char. This raw gas is fed to a secondary gasifier which is essentially an empty vessel. Here, an additional amount of oxygen (and possibly steam) is added. As a result, the temperature increases to typically 1300 °C and the primary product gas together with the entrained carbon particles are gasified virtually completely.

a. **Introduction of feedstock in the bed**

As explained in the report of Creusot Loire, the feeding system consists of two screw feeders in series separated by a rotary valve. The first screw will be cold, the second one can be hot due to contact with the reactor. Because the second screw has a higher feeding rate it will remain almost empty and therefore is not likely to be blocked by pyrolysis products. The rotary valve provides gas tightness to separate the gasifier from the feed bunker. This problem will become more difficult in pressurized operation as foreseen in the future development of the process and probably some additional development will be required.

b. **Danger of ash melting/agglomeration**

This problem has been avoided in the primary reactor by choosing a low operation temperature (± 700 °C), below the ash fusion temperature. In the secondary gasification reactor the temperature is much higher (± 1300 °C) and therefore all the ash will be melted here. In the pilot plant, directly after the secondary gasifier, the product gas is quenched and the ash becomes solid again. It is not completely sure that with this procedure all the ash melting problems have been solved. On what happens with the molten ash particles that may reach the wall no information is presented and in a commercial unit, product gases will probably be cooled down in a heat exchanger where molten ash particles might settle on the cooling surface. Although these problems have been solved in coal gasification processes (Texaco, Shell) their solution has not yet been demonstrated in the Creusot Loire process.

c. **Incomplete gasification, due to a low temperature and/or a short residence time**

Due to the two stage gasification and the high temperature in the second stage, the gasification in the Creusot Loire process is virtually complete.

Consortium or Company	Introduction of the feedstock	Melting;agglomeration of ash	Complete gasification	Solution for entrainment of ash and char	Avoiding expensive oxygen	Safe and effective separation between gasifier and combustor	Heat transfer between beds	Recovery of sensible heat product gas	Prospects of pressurization	Methanol grade gas production	Medium Joule Gas production
Creusot Loire	+	+P ±S	+	+	-	NA	NA	-	+	+	+
Lurgi	+	+	+	+	-	NA	NA	-	±	±	+
John Brown/Wellman	+	±	±	+ ±	±	+	+	±	-	-	+
Italenergie/AGIP	?		?		±		?	?	-	?	?

+ solved/demonstrated
- not solved/demonstrated
± not completely sure
? awaits future testing, only short duration runs available
NA not applicable to this process

TABLE VII Summary of different development items realized in the CEC supported gasification pilot plant project.

d. Methane and other pyrolysis products in the synthesis gas

Due to the high temperature in the second gasification stage there are no pyrolysis products left and the methane content is very low (≈ 0.7%).

e. Entrainment of ash and char particles

Due to the secondary gasification step, practically no char is entrained with the product gas. Ash is removed in a water scrubbing step and thus collected as a water slurry.

f. Use of expensive oxygen

Because of the secondary gasification step, oxygen consumption is relatively high (0.57 kg O_2 per kg wood).

i. Recovery of sensible heat from product gas

No attempt has been made in the pilot plant to simulate these process steps.

j. Prospects for pressurization

No delicate pressurization balances have to be observed inside the plant and on these grounds the prospects for pressurization are good.

Concluding remarks on the Creusot Loire project

Due to a two-step gasification process ash melting problems in the fluid bed are eleminated and a high quality synthesis gas suitable for methanol production has been produced. These features were demonstrated during a 24 hrs testrun. The process shows good prospects with respect to pressurization and this will be demonstrated in the next phase of the project. A disadvantage is the high oxygen consumption. If heat recovery from the product gas has to be applied this has to be demonstrated for this particular process.

2.3 The Lurgi process

Short general description [12]

In the Lurgi process the gasifier is a single reactor with a circulating (fluid) bed (see Fig. 4). It differs from a normal fluid bed by the much higher gas velocities applied, resulting in a high circulation rate of the solids present in the system: reactor - cyclone - feedback pipe. Due to this circulation high temperatures can be maintained in the whole reactor with intensive gas-solids contact. The gasification agent is an oxygen/steam mixture. If the circulating bed can operate without ash agglomeration problems a single step gasification at high temperatures could be realized and producing a synthesis gas low in pyrolysis products.

a. Introduction of feedstock in the bed

The wood is fed to the reactor via a screw feeder and due to the high gas velocities applied in a circulating fluid bed no particular distribution or blocking problems are to be expected.

b. Danger of ash melting/agglomeration

In a testrun, which exceeded 24 hrs, the reactor temperature has been relatively low (' 760 °C) and no ash melting problems have been encountered. At these temperatures gasification was not complete, however, and some pyrolysis products were still present in the product gas (mainly methane). At higher temperatures, where these undesired products for methanol synthesis gas can be eliminated, only short runs have been carried out [13] and therefore the absence of ash agglomeration problems cannot yet be considered to be proven for these conditions.

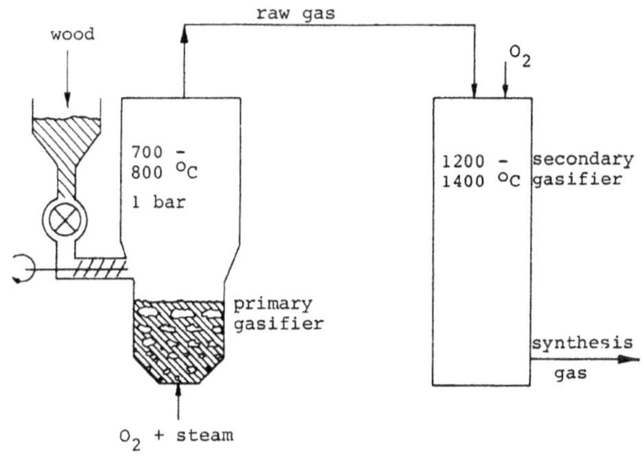

Figure 3 The Creusot Loire Process

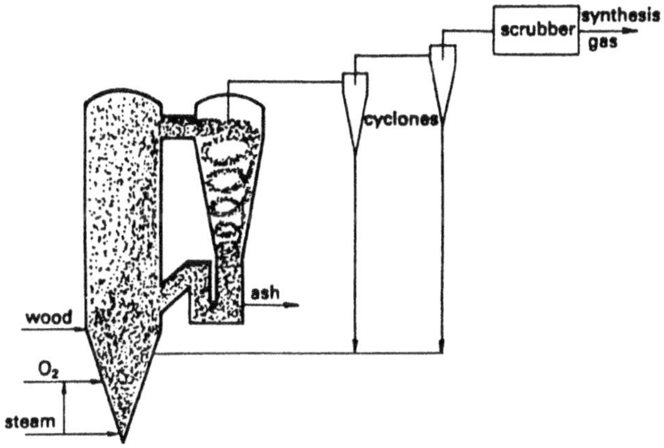

Figure 4 Lurgi's circulating fluidized bed gasifier

c. **Incomplete gasification due to a low temperature**

Considering the temperature level of the testrun, the results show a relatively good conversion of pyrolysis hydrocarbons. Yet the absolute amount (6-7 % vol) of these products is still high for a methanol synthesis gas. The exact value of the carbon/char conversion will also depend on the temperature. The problem of partial conversion can only be solved by increasing the reactor temperature. No experimental results for long duration runs at high temperatures (say 900-1000 °C) are available at the moment. The char conversion efficiency is already good at 760 °C.

d. **Methane and other pyrolysis products in the synthesis gas**

As a consequence of the previously mentioned low reactor temperatures during the testrun the hydrocarbon content of the product gas (mainly methane) is still fairly high (~ 6.5% vol). Just as in the Creusot Loire process this could be solved by secondary gasification (or reforming) but this has not been effected in the Lurgi project. If it is possible to obtain stable operation at a higher temperature a more attractive methanol synthesis gas could be obtained directly. Up till now this has not yet been demonstrated for a sufficiently long period of time.

e. **Entrainment of ash and char particles**

In the circulating bed system entrained ash and char is separated from the product gas by a system of a main circulation cyclone followed by secondary and tertiairy cyclones. Such a set-up indeed may allow for a recovery of ash/char mixtures at different positions in the separation process, possibly with varying char content.

f. **Use of expensive oxygen**

During the long duration testrun, the oxygen consumption was 0.45 kg/kg dry wood. This seems to be significantly lower than reported for the Creusot Loire process (0.57 kg/kg dry wood) but it should be realized that the product gas in the Lurgi process still contains an important quantity of hydrocarbons. Perhaps the oxygen consumption should also be compared with the first gasification step of the Creusot Loire process (0.2 kg/kg dry wood).

i. **Recovery of sensible heat from product gas**

This process step has not yet been piloted, and inclusion in a commercial plant has not yet been stated. Problems could be fouling of the heat exchanger due to remaining hydrocarbons (tars) and ash/char dust.

j. **Prospects for pressurization**

Lurgi did not announce a pressurized version of the process and because the circulating fluid bed concept still has a relatively short history, little can be said about the possibilities of pressurization.

Concluding remarks on the Lurgi project

The circulating fluid bed gasifier operated satisfactorily and produced, during a relatively long duration run, a medium Joule product gas. On a pilot scale, methanol was produced from this gas but the hydrocarbon content (mainly methane) of the gas was still on the high side. In principle, at higher reactor temperatures (say 900-1000 °C) it should be possible to obtain a gas nearly free from hydrocarbons possibly even at higher capacities. Unfortunately, up till now it was not demonstrated under such conditions for a longer period of time. Possible difficulties might be unstable operation due to low char hold-up at high temperatures and/or ash agglomeration or melting.

2.4 Oxygen Donor Gasification of John Brown/Wellman

Short description of the process [14]

The process produces a medium Joule gas using biomass and air only. The reactor consists of a double fluid bed (see fig. 5) with intensive exchange of bed material between the two bed compartments via a special slot. Part of the bed material acts as an oxygen carrier from oxidiser to gasifier. In the oxidiser, oxygen is chemically bound by calcium sulphide according to:

$$CaS + Air \rightarrow CaSO_4$$

Additionally, also some char present in the bed material is combusted in the oxidiser to provide additional heat to the gasifier system. The relatively hot, sulphate rich bed material flows from the oxidiser to the gasifier where oxygen is recovered by reduction of the bed material with the gases present in that reactor:

$$CaSO_4 + 4H_2 \rightarrow CaS + 4H_2O$$

$$CaSO_4 + 4CO \rightarrow CaS + 4CO_2$$

These reactions provide the gasification reactants H_2O and CO_2 which, after reaction with the biomass pyrolysis products, produces a surplus amount of carbon monoxide and hydrogen.

With an additional amount of steam injected in the gasifier section and replacing the $CaSO_4/CaS$ carrier by inert material the system could in principle also operate as double bed steam gasifier.

a. **Introduction of feedstock in the bed**

The gasifier is completely surrounded by the oxidiser [14]. Therefore, the biomass feedstock has to be introduced either through the bottom or through the top of the reactor. By a specially designed pneumatic feeder in series with a two screw feeder sluice system this problem has been solved successfully.

b. **Danger of ash melting/agglomeration**

This problem should be especially important for the regenerator section where the highest temperatures are found (up to 1000 °C). Although a good circulation behaviour was observed in the hot bed, ash agglomeration still may become visible after a prolonged operation time. Moreover, ash agglomeration may depend on the stone refreshing (consumption) rate in the fluid bed which was very high during the particular test reported by John Brown/Wellman. Therefore, though ash agglomeration problems have not been observed yet, long duration tests are still necessary for a final proof on this aspect.

c. **Incomplete gasification, due to a low temperature and/or a short residence time**

The data reported by John Brown/Wellman do not allow for a calculation of the char conversion, specially not because extra coal was added to the oxidiser. However, due to the coupling with a combustor, there is no reason to doubt the char conversion to be insufficient. It may be concluded that the product distribution is similar to that of a double bed steam gasifier. In fact, the evidence that the system worked in the oxygen donor mode is based on the strong temperature dependence of $CO + H_2$ formation in the gasifier and on the fact that these gases were produced at a small partial water-vapour pressure in the inlet gas of the gasifier. In any case, it can be safely concluded that to reach the CEC bonus set of conditions a secondary gasification or reforming step will be necessary.

d. Methane and other pyrolysis products in the synthesis gas

The product gas of the ODG still contains a (too) large amount of hydrocarbons, reportedly nearly all methane (12-13 vol %). The contractors are right in stating that the process has not been optimized yet, so that this figure still may go down as a result of future optimization programme. On the other hand, we have always been sceptical on predictions with respect to methane which were substantially below those for steam gasification in fluidized beds at corresponding temperatures.

e. Entrainment of ash and char particles

Ash and char are removed by cyclones, preferably downstream of the oxidiser because here the ash contains the smallest amount of char. In principle, the char content in the ash can be low in a double fluid bed system, but the ODG process may require a minimum of char present in the oxidiser to prevent SO_2 emission via the regenerator exhaust gases. Because also some coal was fired in the regenerator during the biomass test, no quantitative data on char entrainment can be given nor on char conversion. A special complicating factor with the ODG process is the stone handling.

f. Use of expensive oxygen

No pure oxygen is used in the process. The methane content, however, is still too high. May be, the contractors will succeed in getting this figure sufficiently reduced by optimizing the process in a follow-up programme. But if this fails and if secondary gasification has to be applied for getting a synthesis gas sufficiently low in hydrocarbons, then, still substantial amounts of oxygen will be necessary.

g. Safe and effective separation between gasifier and combustor

Product gas composition indicates a sufficiently effective and probably safe separation between oxidiser and gasifier.

h. Heat transfer between gasifier and combustor

Both from direct measurements in cold models and from results during gasification it can be concluded that the heat transfer with the solids flow through the slots is more than sufficient and relatively simple to operate.

i. Recovery of sensible heat

In the pilot plant an air preheater was installed which recovered sensible heat from the regenerator flue gas. Some difficulties were encountered during the operation due to entrained fines that slipped through the cyclone system. Although these problems may easily be overcome in further developments it demonstrates once more that for complex processes like wood gasification a completely integrated pilot plant is required to be sure that any surprise with "standard technology" in full scale plants will be eliminated.

j. Prospects for pressurization

Due to the complexity of the process it does not seem logical to pressurize the process and no developments towards a pressurized version are foreseen.

Concluding remarks on the O.D.G. process of John Brown/Wellman

The pilot plant has been constructed and operated for several runs of which the longest reportedly lasted for 9.5 hrs. A medium Joule value gas was produced without the use of pure oxygen. The product gas did not fulfill the desired methanol synthesis gas composition as mentioned in the bonus conditions (see Table III). The methane concentration was still too high. Without the introduction of a secondary gasifier or reformer it is unlikely that the methane content of the product gas can be sufficiently reduced. It is not completely sure yet that the plant indeed produced its synthesis gas mainly via the Oxygen Donor principle as a certain fraction of the gas may have been produced by other mechanisms like steam gasification in a double bed. Unfortunately pure steam gasification which easily could have been tested with (nearly) the same equipment was not carried out. At the present state of affairs the stone consumption is still rather high and should be reduced to make the process attractive. A complicating factor is that during operation on biomass, always some coal was co-fired in the combustor and no data on pure biomass conversion are available yet.

Another uncertain point is the potential emission of traces of SO_2 from the oxidiser. This has not been reported but no data on the actual value of SO_2 in the oxidiser flue gas have been published either. If the flue gas contains a substantial amount of SO_2 originating from $CaSO_4$ than this would be a particular disadvantage for wood which is a clean almost sulphur free feedstock. Nevertheless, in applications like power generation or combined heat/power generation the ODG process looks attractive both for biomass and coal as feedstocks. The question remains whether the lower or zero steam consumption justifies the rather complex chemical system relative to the simple alternative of steam gasification with inert heat carrier in almost the same equipment, particularly if the feedstock is biomass. For methanol production, the product gas still should be subjected to secondary gasification or steam reforming unless a substantial improvement in operating performance will be realized during a follow-up programme.

2.5 <u>The Italenergie/AGIP-S.p.A. process</u>

Short description of the process [15]

A simplified scheme of the process is given in fig. 6. A medium Joule gas is produced by steam gasification of wood in the inner fluid bed. The heat for this endothermal process is provided through the corrigated wall of the reactor. Part of the gas produced is combusted in a second fluid bed located in the annular space of a vessel around the gasifier. To produce a methanol grade gas, the product gas of the steam gasification is further converted in a secondary gasifier (cracking chamber) where oxygen is introduced, raising the temperature to 1300 °C. Upon completion of their report the plant had not yet been in full operation for runs of more than two hours and was still in the initial testing phase.

a. **Introduction of feedstock in the bed**

Because the feedstock has to be introduced into the central bed, just as in the case of John Brown/Wellman, this is a specially complicated matter and the technical solution will have to show its usability.

b. **Danger of ash melting/agglomeration**

No problems with ash melting are to be expected in the gasifier because of the relatively low temperature ($T \approx 730$ °C). Ash melting in the secondary gasifier is a potential problem only in case wood ash traces can be transported to this vessel. In the present set-up, the danger of ash melting is low in the combustion chamber because the fuel used here is gas (with possible traces of ash/char) and the temperature is relatively low (850 °C).

c. **Incomplete gasification due to a low temperature and/or a short residence time**

Due to the relatively low temperature in the gasifier it should be

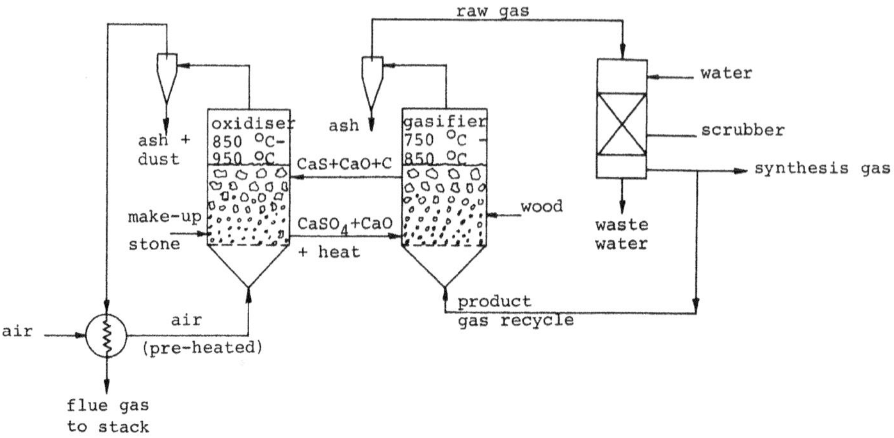

Figure 5 The Oxygen Donor Process of John Brown/Wellman

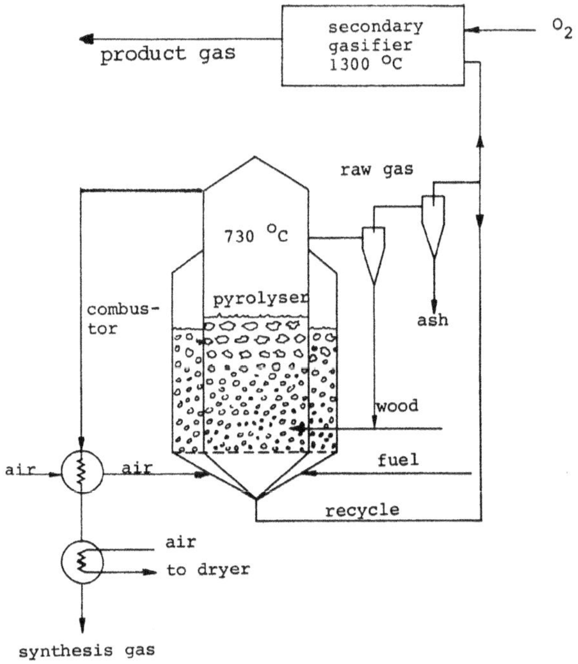

Figure 6 The Italenergie/AGIP S.p.A. process

doubted that a full char conversion can be realized, specially not with steam gasification. The primary product gas will contain still a lot of hydrocarbons and methane but for methanol synthesis gas production this gas is sent to the secondary gasifier where these products are converted.

d. **Methane and other pyrolysis products in the synthesis gas**
No problems here because the secondary gasifier is expected to convert these products to CO and H_2.

e. **Entrainment of ash and char particles**
The entrained ash and char particles in the primary product gas have to be removed at a high gas temperature. Fractional slip of ash particles may lead to blocking and fouling of the gas recycle to the gasifier, to ash entrainment to the heater/combustor and to ash melting problems in the secondary gasifier. Long duration runs with the pilot plant will show whether these potential problems can be avoided.

f. **Use of expensive oxygen**
No oxygen is consumed in the primary gasification step. In the secondary gasifier oxygen will be consumed and based on the Creusot Loire experience the quantity is expected to be considerable.

g. **Safe and effective separation between gasifier and combustor**
Several potential problems can be identified for the corrugated separation wall between gasifier and combustor such as maintaining gas tightness during thermal expansion, erosion/corrosion, material stresses etc. A hot wall between two fluid beds is a relatively new development item on which hardly any experience is available. Also here, only the pilot plant test of sufficient duration will indicate whether the technical problems have been solved.

h. **Heat transfer between gasifier and combustor**
Because the gasifier is heated-up indirectly through a corrugated wall and because fluid beds can provide high heat transfer coefficients no problems are expected with the heat transfer itself. On possible material problems with the heat exchanging wall; see under g.

i. **Recovery of sensible heat from product gas**
Heat exchangers are provided in the pilot plant to recover sensible heat from the product gas and from the combustor flue gases but we will have to wait for operational data to judge the performance.

j. **Prospects for pressurization**
Because of mechanical problems to be expected it seems to be unlikely that the heat exchanging wall will coincide with the separation of the pressurized/non pressurized part of the process. Therefore the whole reactor system (combustor/gasifier/secondary gasifier) will have to be pressurized. This is not impossible but certainly rather difficult.

Concluding remarks on the Intalenergie/AGIP-S.p.A. process
Indirect heating of the gasifier through a wall seems to be an attractive route to produce raw synthesis gas if it can be realized on a technical scale at a sufficient rate and degree of biomass conversion. If secondary gasification is done with oxygen, much of the attractiveness is lost as a large amount of oxygen will still be used. However, for medium Joule gas production and with a (secondary) steam reforming step, also in the case of methanol production, the process regains much interest. A critical condition

will still be whether gas clean up from tar and particles will be sufficient for these cases.

2.6 Additional pilot projects

Two additional pilot projects have been incorporated in the present pilot programme of the CEC:
- The AVSA fluid bed Combustor-Gasifier Project
- The Twente University of Technology project on hydrogen recovery.

2.6.1 **The AVSA fluid bed Combustor-Gasifier Project (University of Brussels)**

In this project [16] a double fluid bed gasifier has been developed with a simple but ingeneous system of solids circulation (see figure 7). To circulate the solids between gasifier and heater and vice versa the solids are transported from a fluid bed with a lower gas velocity (and therefore higher density) via a hole or slit to a fluid bed with a higher gas velocity (and lower density). The driving force for solids transport is the difference in pressure on both sides of the holes. Both the gasifier and the combustor each consist of two beds separated by a wall, one with a relatively high, the other with a relatively low gas velocity. Solids are transported from the bed with a high gas velocity to the bed with a low gas velocity by flowing over the top of the wall separating these two sections. Very high solids transport rates can be obtained (up to 1000 kg/per m^2 transport hole area). Also the important problem of sufficiently large biomass solids circulation and segregation has been studied and solved.

A pilot unit has been constructed, but up to now only the combustion section has been tested successfully. Present problems are the gas exchange rate between gasifier and combustor which is too high (20-40%) and more development will be required to reduce this quantity. Another problem will certainly be the high hydrocarbon content of the product gas (as can be derived from small scale experiments). For methanol production this is an important disadvantage but for many other applications it will often be acceptable, provided the tar content is low. On the latter aspect, no reliable data are available yet. Char conversion can be relatively high in a double bed gasifier due to the combustion section but tar/char/ash dust mixtures entrained with the product gas may still form a difficult problem if a clean gas is desired. Nevertheless, the project forms an interesting addition to the CEC program, extending the value and being complementary to the John Brown/Wellman and the Italenergie/AGIP projects.

2.6.2 **The Twente University of Technology project on hydrogen recovery**

This project [17] aims at the selective absorption of hydrogen at low partial pressures typical for low Joule producer gas obtained from simple atmospheric air gasification of wood. With slurries containing finely dispersed hydridible metals it is possible to recover the hydrogen continuously and get it available in a pure form at increased pressures. Because continuous processes to recover CO already exist, it should be possible this way to produce pressurized synthesis gas e.g. for methanol production from low Joule gases obtained from simple air gasification. By steam shifting of all CO to H_2 it is also possible to produce hydrogen only, which then could be used e.g. for hydrogenation or biomass liquefaction. Presently, only small scale laboratory tests are available but they look quite promising. Also a continuous pilot plant (fig. 8) including the hydrogen absorber and desorber has been constructed and is being started up at this moment. Important development items are prevention of poisoning of the metal alloy slurry and possibly still increasing the volumetric absorption rate. The process could

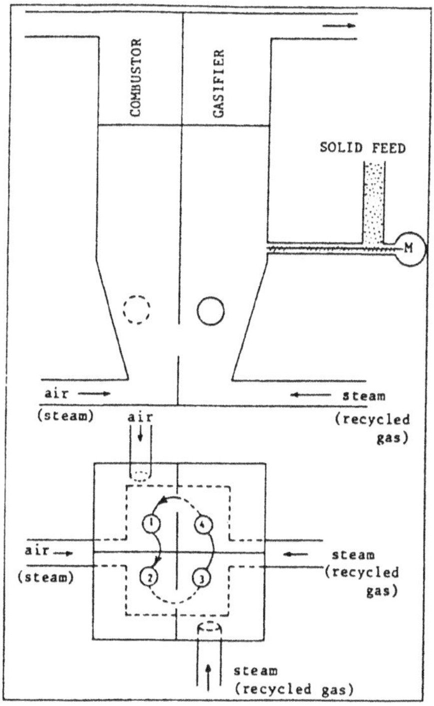

Figure 7 Schematic view of the AVSA process (University of Brussels)

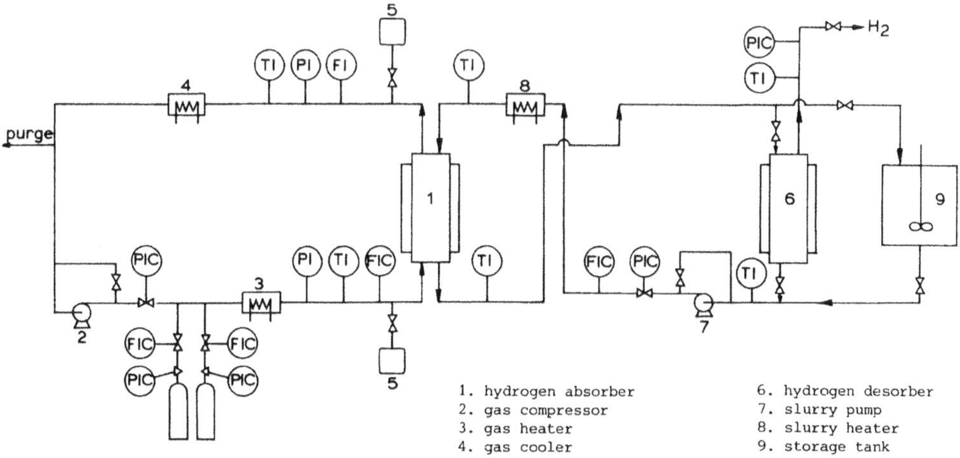

Figure 8 Experimental hydrogen separation and recovery unit (Twente University of Technology)

play an important role in several kinds of biomass conversion processes if it can be developed up to a commercial scale.

3. FUTURE DEVELOPMENTS

The four gasification pilot plant projects together with the additional pilot projects form an interesting platform for future developments both for future methanol from wood processes as well as for other biomass to energy processes.

For the methanol route it is important that developments will allow a methanol plant to be economically feasible on a 100.000 ton/year scale producing fuel grade methanol. This will require several modifications on the present processes which are directed to high purity methanol for chemicals on a 1.000.000 tons/year scale. Probably the new fuel grade methanol and the new liquid phase methanol processes are good candidates for these developments.

Another factor is the production of a cheap synthesis gas. Avoiding the use of expensive oxygen resulted in the present pilot programme always to synthesis gases rich in hydrocarbons. Therefore, it would be interesting to develop or to adapt steam reforming processes which could convert these gases to better synthesis gases; this should preferably be done immediately after the primary gasification step.

Another way out could be air gasification (primary and secondary) and recovery of the synthesis gas components from the low Joule gas.

Apart from possible methanol synthesis, the processes have a potential application in several other fields like: production of reducing gas, fuel gas, electricity with gas turbines, diesel aggregates or fuel cells and hydrogenative liquefaction of biomass.

As stated in their paper the Creusot Loire project will be followed up by a much larger pilot project and also other projects may be continued in some form.

Now, in 1985, particularly on the short term, the economic incentives for large scale synthesis gas production from biomass do not look very bright, at least not within the European Community. On the long term, however, it might well become the most prospective and attractive way of harvesting solar energy. The pilot programme of the CEC has certainly given an important contribution to this development.

REFERENCES

1. S. Hales, Vegetable statics or an account of some statical experiments on the sap in vegetables, being an essay towards a natural history of vegetation of use to those who are curious in the culture and improvement of gardening. Also a specimen of an attempt to analyse the air, by a great variety of chemico statistical experiments, which were read at several meetings before the Royal Society, London (1727). Reference found in: A.J.J. van de Velde, Jan Pieter Mickelers en het steenkool gas, Mededeling van de Koninklijke Vlaamse Academie van Wetenschappen, Letteren en Schone Kunsten van Belgie, Brussels (1948).

2. A.A.C.M. Beenackers and W.P.M. van Swaaij, Methanol from Wood I+II, Int. J. of Solar Energy, (1984) 2, 349 and 519.

3. A.A.C.M. Beenackers and W.P.M. van Swaaij, Gasification of Biomass, A State of the Art Review, in: A.V. Bridgwater (Ed.), Thermochemical Processing of Biomass, Butterworths (1984) London, 91-136.

4. T.B. Reed, A Survey of Biomass Gasification, Vol III, SERI, Golden (1980); SERI/TR-33-239-VI.

5. Ullmann's Encyclopädie der technischen Chemie, 4th Ed., Band 16 (1978) 624.

6. H. Hiller, Co-production of methanol and higher alcohols for automotive uses, E. supp, European Methanol Conf., Dec. 1984.

7. M.E. Frank, in: Proc. Intersoc. Energy Conversion Eng. Conf. 15th (2) (1980) 1567.

8. D.M. Brown, in: ISCRE-8, The Inst. Chem. Engrs Symp. Ser. no 87 (1984) 699.

9. Call for tenders, Publikatie blad van de Europese Gemeenschappen Nr C 18/2 d.d. 27-1-1981.

10. W. Palz and G. Grassi (eds.), Energy from Biomass, Vol 2, Proceedings of the Workshop on Biomass Pilot Projects on Methanol Production and Algae, held in Brussels, 22 October 1981, Reidel, Dordrecht (1982).

11. Fluidized bed oxygen gasification of wood G. Chrysostome and J.M. Le Marle, Framatome, Division Cleulot Energie, Le Creusot, France.
Final project report (to be published).

12. Synthetic fuel from wood via gasification in the circulating fluid bed.
P. Mehrling and R. Reimert, Lurgi GmbH, Frankfurt/Main, F.R. Germany.
Final report project (this volume).

13. P. Mehrling, pers. communication.

14. The gasification of wood using the Oxygen Donor Process.
I. Dallas, R.J. McLellan, K.E. Scanlon and D.H. Smith.
John Brown engineers and Constructors Ltd, London and Wellman Mechanical Engineering Ltd, Smethwick, Warley, West Midlands, United Kingdom.
Final report project (this volume).

15. Gasification of biomass for the production of synthesis gas with the intention to produce synthtic fuel in 2 further process.
G. Coglisti Italenergie S.p.A., Sulmona and AGIP S.p.A., Roma, Italy.
Final report project (this volume).

16. The AVSA fluid bed combustor-gasifier project, H.A. Masson.
University of Brussels, Belgium.
Final report project (this volume).

17. Hydrogen recovery from gas mixtures using metal hydrides suspended in slurry.
K.J. Ptasinsky, A.A.C.M. Beenackers, W.P.M. van Swaaij and R. Holstvoogd.
Twente University of Technology, Enschede, The Netherlands.
Final report project (this volume).

FINAL REPORTS

FLUIDIZED BED OXYGEN GASIFICATION OF WOOD

Authors : G. CHRYSOSTOME - J.M. LEMASLE

Contract number : ESE-P-068-F

Duration : 21 months

Total budget : 8 500 000 FF C.E.C. contribution : 4 000 000 FF

Head of project : G. CHRYSOSTOME

Contractor's adress : FRAMATOME Division CREUSOT-ENERGIE
BP 13
71201 LE CREUSOT CEDEX - FRANCE

SUMMARY

The "New Products" department of FRAMATOME Division CREUSOT-ENERGIE has been involved in the "methanol from wood" R and D program of the C.E.C. (and the French A.F.M.E.) since 1980.
The purpose of that program is to develop wood gasification processes for syngas production.
In the Laboratory of Energetic Testings of Le Creusot, a pilot unit was built at the end of 1980, on the principle of the fluidized bed wood gasification with oxygen and steam.
The research developed during the past four years led to the conception of a two steps process. The first one is composed of a fluidized bed gasifier in which wood chips are continuously converted in a steam and oxygen stream at 700-800°C. The gas leaving that first stage of gasification mainly contains CO, H2, CO2 and H2O but also significant amounts of CH4 and higher hydrocarbons ; such a gas is not suitable for methanol synthesis. In the second step, the raw gas leaving the fluidized bed gasifier reacts with additional oxygen into a partial oxidation reactor (called "secondary reformer") ; methane and higher hydrocarbons are cracked at about 1 300°C and converted into CO, H2, CO2 and H2O.
The atmospheric pilot has been operated at Le Creusot since 1981, at a wood capacity of 100-200 kg/h at the beginning, then 400 kg/h since the end of 1983. The secondary reformer has been operated since July 1982. Long duration runs since the end of 1982 have confirmed expected results in terms of methanol yield (a potential production of 0.5 kg methanol per kg dry wood gasified has been experimentally achieved).
A 24 hours "guarantee run" has been performed, at the beginning of 1984, to fulfill the contract between CREUSOT-ENERGIE and the C.E.C. During that continuous run, a capacity of 10 tons per day was demonstrated and the main gas characteristics were CO + H2 > 70 % and total inerts content of syngas < 3 %.
A pressurized pilot unit based on the same process will be implemented at Clamecy (department of Nievre - France) with the financial help of the A.F.M.E. (French Agency for Energy Savings) and of the C.E.C. (Commission of the European Communities). The nominal capacity of this unit will be 60 tons per day (dry basis) of wood under a 15 bar nominal pressure. Preliminary testings of the pilot plant of Clamecy will begin in early 1986.
For that research, CREUSOT-ENERGIE has been working in collaboration with the University of Technology of Compiegne (U.T.C.) which provided some support in the study of the pyrolysis of wood, in elaborating gasification models and, at the beginning, in the gas analysis.

TABLE OF CONTENTS

1 INTRODUCTION

To present a general overview of the work which has been performed in that "syngas from wood" program, we hereunder have drawn the main steps of the research :

1.1 Basic research on the pilot

1980 : Construction of a fluidized bed gasifier, 400 mm in-diameter, atmospheric pressure

1981 : Experimentation of the pilot - Short runs (up to 5 hours each)

1982 : Construction of a secondary oxygen reformer, 500 mm in-diameter Short runs with the two stages pilot

End 1982 - mid 1983 : 3 long duration runs with the two stages atmospheric pilot plant (each test : 100 hours) - Wood capacity ≃ 100 kg/h

Mid 1983 - end 1983 : modification of the atmospheric pilot to rise its capacity up to 10 tons/day of wood

- new fluidized bed gasifier : 500 mm in-diameter
- new secondary reformer : 800 mm in-diameter
- now wood feeding system...

Begin 1984 : 24 hours guarantee run for the C.E.C.

- 10 tons/day of wood gasified
- less than 3 % total inert content of the syngas
- $CO + H_2 > 70$ %

Filling of bottles for test of methanol synthesis at LURGI.

1.2 Parallel work

. Atmospheric pyrolysis and gasification of wood pieces in an atmospheric thermal balance → results on kinetics of thermal degradation of wood (University of Technology of Compiegne)

. Basic engineering of a 60 TPD wood gasification unit under pressure - Pilot of Clamecy

2 DEFINITION OF THE PROGRAM

The program of the research is defined in the contract n° ESE-P-068-F between the Commission of the European Communities and CREUSOT-ENERGIE.

The main terms of the contract are :

- CREUSOT-ENERGIE shall be responsible for the modification of an existing gasification reactor for gasification trials with a throughput of 10 tons per day of wood to obtain a synthesis gas suitable for methanol production. The work shall be completed by 31 March 1984.
- The commission shall bear 50 % of the actual cost up to a maximum of FF 4 000 000, excluding all taxes. The Commission will increase its participation up to 10 % of the actual cost up to a maximum of FF 4 800 000 provided that the installation is modified before 31 March 1984 and gas up to the specifications, has been produced and delivered for methanol synthesis demonstration tests.

* Conditions to obtain the 20 % bonus

The syngas leaving the installation should meet the following requirements :

- Syngas samples will be taken at regular intervals during a 24 hours period of gasification. These samples will be taken in the presence of experts of the C.E.C.
- Samples of syngas should contain :
  . at least 70 % volume of hydrogen plus carbon monoxide
  . less than 3 % volume of inerts (nitrogen + methane + higher uncondensable hydrocarbons).
- The installation should gasify 400 kg/h of wood, 20 % moisture.

* Time schedule of the work

Activity	end - date
. Begin of the project	01/07/82
. Construction of a secondary reformer	01/09/82
. Preliminary testings of the secondary reformer	31/10/82
. 3 long duration runs (100 hours)	15/04/83
. Studies for rising the capacity of the atmospheric pilot ut to 10 tons/day	31/07/83
. Beginning of modifications on the pilot	15/09/83
. End of modifications on the pilot	30/11/83
. Begin of gasification tests on the modified pilot	01/12/83
. Guarantee run of 24 hours in the presence of experts of the C.E.C.	15/02/84
. Delivery of gas samples for methanol synthesis	28/02/84

3 DESCRIPTION OF THE ATMOSPHERIC PILOT PLANT OF LE CREUSOT

The pilot is represented on the following figures 1 and 2.

The pilot (see schematic diagram on figure 2) is mainly composed of a 0.50 m in-diameter fluidized bed gasifier and a 0.80 m in-diameter secondary reformer. The gasifier is composed of a steel casing with an outside insulation. At the bottom, a fluidizing grid ensures the injection of an oxygen-steam mixture. The height of the gasifier is about 5 m. The height of the conical part at the bottom is 0.6 m. In the upper part of the gasifier, a 0.7 m in-diameter disengaging height allows lower velocities of gases and so lower fine particles elutriation.

Wood chips (0.05 x 0.02 x 0.01 m max) are fed with the help of a semi-continuous feeding system into a 0.7 m^3 charge hopper, then fed into a variable rotating screw feeder for adjustable feed-rate. Woods chips then fall dawn into a rotary valve and finally are fed into a second screw feeder at a high rotating rate for rapid transfer towards the hot zone of the reactor.

The conical bottom of the reactor is filled with an inert material (sand) which is the fluidizing material. Such a fluidized bed allows to handle biomass of a rather wide range of size distribution (up to 50 mm).

At the very beginning of each test run, preheating of reactors is re-formed by natural gas combustion with air injected at the bottom of the fluidized bed, and at the top of the secondary reformer. So, temperatures of 800°C in the gasifier and 1 200°C in the secondary reformer are reached.

The gasifier is equipped all over its height with 18 thermocouples (chromel alumel). Pressure taps control the pressure drop in the fluidized bed. The pressure above the fluidized bed and in the overall plant is regulated just below the atmospheric pressure to avoid hazardous gas emissions in the environment of the pilot.

A small part of gases leaving the gasifier is directed to the analysers. The dry gas composition is measured with an infrared analyser for CO and CO2 (continuously), with a paramagnetic analyser for O2 (continuously) and with a gas chromatograph (GIRDEL) for CH4, C2H4, C2H6, H2, N2, Ar. An argon tracer is injected, at a controlled flow-rate in the gasifier, allowing to determine the dry gas flow-rate produced by the installation and so, to perform mass and energy balances.

The gas leaving the fluidized bed gasifier (between 600 and 700°C) then enters into the secondary reformer where secondary oxygen is also injected. The secondary reformer is a 5 m height vertical cylinder, refractory lined, allowing to work at temperatures greater than 1 300°C. Temperatures in that reactor are measured with 6 thermocouples (Pt - Pt.Rh).

As at the outlet of the fluid bed gasifier, a small part of gases leaving the secondary reformer is also derived to the analysers. Every 10 minutes, a complete gas analysis from the gas chromatograph is obtained.

The main stream of gas leaving the secondary reformer is then cooled into a water scrubber (0.6 m internal diameter, 2 m height), sucked up by an exhauster and finally incinerated at the flare.

All flow rates of various fluids are measured by rotameters. The wood feed rate is measured by accurate weighing of each charge fed into the hopper. Its moisture is determined by drying of samples and then weighing them.

Spent water can also be analyzed to determine its composition.

In this configuration, the capacity of the pilot is about 400 kg/h of wood.

In its previous configuration (up to June 1983), the atmospheric pilot of Le Creusot had a 150 to 200 kg/h capacity. It was mainly composed of a 0.4 m in-diameter fluidized bed reactor, a 0.5 m in-diameter secondary reformer and a 0.8 m in-diameter post combustion chamber. The previous wood-feeding system was simply composed of a pre-charge hopper at the bottom of which an endless screw was feeding wood chips into the gasifier, at a controlled feed-rate. With such a simple system, uncontrolled air was entering into the gasifier, by the endless screw, due to the working conditions just below the atmospheric pressure. So, the gas composition was greatly affected by a 10 to 20 % nitrogen content, and we were obliged to check the corrected composition of the produced syngas.

The safety measures and equipment of the atmospheric pilot plant are :
-ruptury disk on the gasifier,
- external CO detectors in the direct environment of the gasifier, with alarms at 50 ppm and 200 ppm,
- alarms on the pilot burners in case of flame extinction during the pre-heating of gasifier and secondary reformer,
- automatic shut-off of oxygen flow rate with automatic nitrogen injection in several cases (power-cut, high pressure of oxygen, failure in the product gas exhauster, failure in the wood feeding system).

FLUIDIZED BED WOOD OXYGEN GASIFICATION
ATMOSPHERIC PILOT PLANT OF LE CREUSOT

FIGURE 1

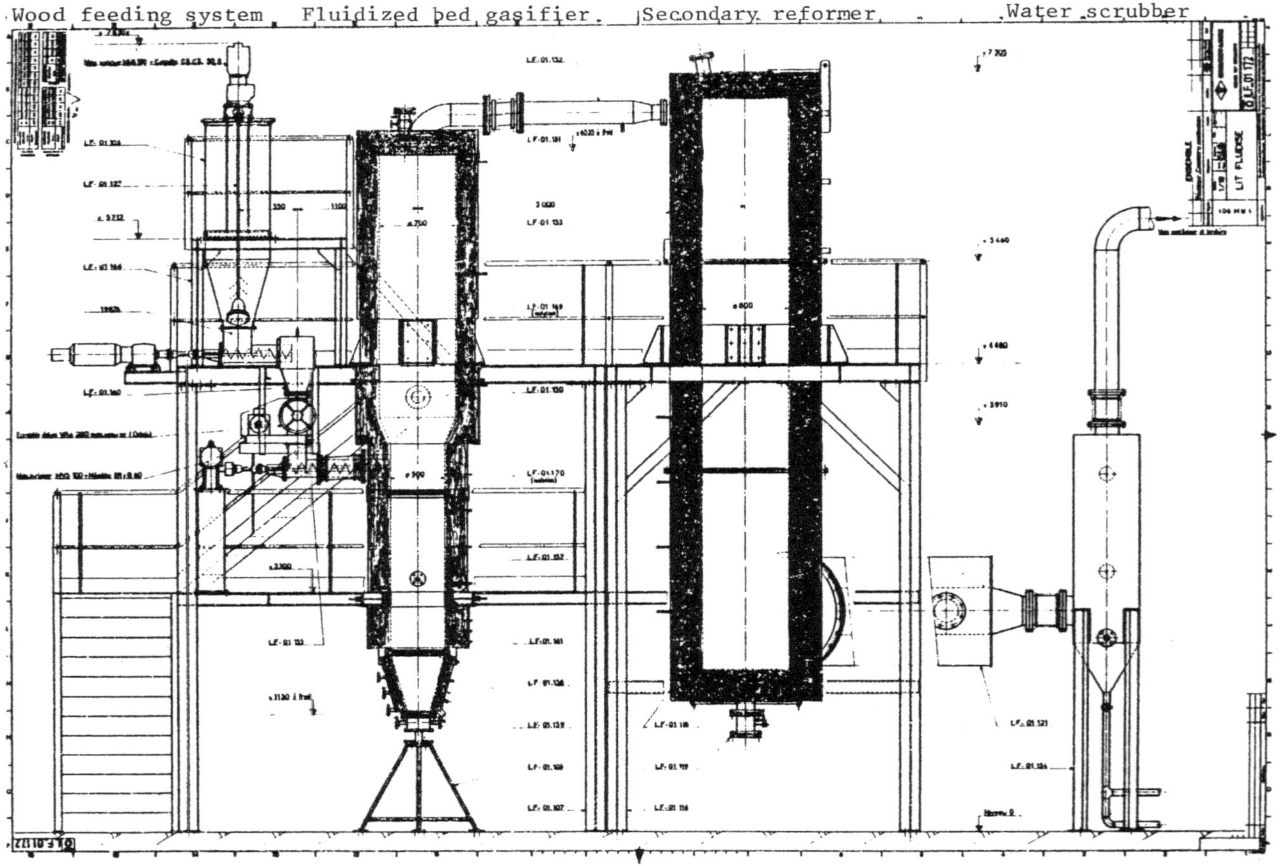

SCHEMATIC DIAGRAM OF THE ATMOSPHERIC PILOT PLANT OF LE CREUSOT

FIGURE 2

4 PYROLYSIS AND GASIFICATION

4.1 Properties of wood

Wood is a solid fuel having a high volatile matter (70-80 %) and oxygen content.

For a dry-ash-free wood, the composition is approximately :

- C : 50 %
- H : 6 %
- O : 43.5 %
- N : 0.5 %

and the global molecular formula can be : $C_6H_9O_4$. In the dry-ash-free state, the low heating value ranges between 4 200 and 4 500 kcal/kg, depending on the wood species. Our material and heat balances in the further chapters are based on a standard heat of formation for wood of -188.4 kcal/mole at 25°C, a low heating value of 4 386 kcal/kg dry-ash-free.

Wood which has been naturally dried in the ambiant air stabilises its moisture content at about 20 %. When wood is heated, it primary delivers water vapor, then above 275°C fuel gases which spontaneously burn in air between 450 and 650°C (pyrolysis). After that pyrolysis, the residual carbon represents only 15 to 30 wt % of the initial wood.

4.2 Ashes and their composition

The combustion in air of a char sample gives some ashes that contain mineral elements coming from the product of origin. In the following table is given the proportion and composition of ashes depending on the different parts of the tree.

	% ash/dry wood	% in ashes			
		K_2O	P_2O_5	CaO	MgO
Green bark	5.0	30	15	30	8
Leaf	4.5	25	12	35	8
Ripe bark	2.0	20	12	50	6
Branch	2.0	10	5	55	6
Soft wood	0.4	10	4	57	5
Hard wood	0.25	11	3	60	5

Note : x % K_2O, P_2O_5, CaO or MgO doesn't mean that ash contains K, P, Ca, Mg under the form of oxides but that K, P, Ca, Mg converted, as an assumption, in oxides represent x % of the ashes.

Depending on the different species, the ash fusion temperature can vary greatly. Some experiments performed at CREUSOT-ENERGIE on hard wood lead to a fusion temperature over 1 400°C.

Experiments made at I.G.T. gave the following results :
ash fusion temperatures, °C

	Douglas fir		Maple hardwood	
	Reducing (°C)	Oxidizing (°C)	Reducing (°C)	Oxidizing (°C)
Initial Def. IT	1 132	1 177	> 1 480	> 1 480
Fluid. FT	1 354	1 463	> 1 480	> 1 480

Note : These normalized terms are resulting of the testing procedure, during which several phenomenous are observed. First of all, ashes are gathered to take the shape of a cone ; then ashes are heated.

- The initial deformation temperature (IT) is reached when the initially sharp top of the cone becomes quite rounded.
- The fluid temperature (FT) is reached when the cone has become completely flat (ashes fully melted).

In the fluidized bed made of an inert material, some interactions between wood ashes and the chemical compounds of the "inert" material can lead to agglomerating conditions. This is the reason why we are obliged to work at rather low temperatures in the fluidized bed gasifier (700-800°C).

4.3 Pyrolysis of large wood samples (atmospheric pressure)

That part of the work has been performed at the University of Technology of Compiegne.

Wood gasification proceeds through a complex mechanism, involving simultaneous and consecutive physical and chemical transformations. After a drying step, pyrolysis occurs, releasing gases, tars and charcoal. Then the reaction of steam with the carbon leads to synthesis gas and heat is supplied by the partial combustion of carbonaceous materials with oxygen. The rate of the thermal degradation, as well as the nature and quantities of the various products, depend, to a large extent, on the temperature. Because of the heterogeneity of the reacting system, the overall kinetics depend also on the size of the wood particles. So, to properly design a wood gasifier, it would be necessary to understand the kinetics of the different steps and the interaction between the chemical reactions and the heat and mass transport phenomena.

Most of the published results about wood pyrolysis have limited applicability to real gasification situations. We here provide some information about the pyrolysis of wood chips and develop a simple model which predict the time required to complete the conversion when the temperatures varies between 500°C and 900°C. We have designed a simple experimental system, able to handle large wood samples, in order to obtain the sample weight loss as a function of time. Experimental results have enabled us to build a simple mathematical model, which includes heat conduction, with a variable thermal conductivity.

4.2.1 The experimental system and results

The kinetics measurements were carried out with a simple system, which provides the weight variation of slabs or of cylinders with diameters up to 27 mm when heated to temperatures as high as 900°C.

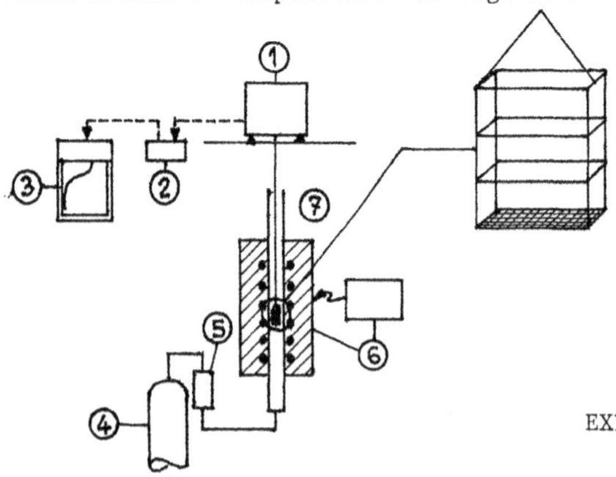

1. Precision scale
2. Converter
3. Recorder
4. Comprimed gas
5. Flow-meter
6. Oven + regulation
7. Quartz reactor

EXPERIMENTAL DEVICE

FIGURE 3

Experimental system and procedure

This device, illustrated in Figure 3, consisted of a vertical tubular kiln, containing a quartz tube. The temperature inside the chamber is controlled and recorded, and a gaseous mixture, with a fixed composition, passes upward through the tube and its rate is monitored.

The wood sample is placed in a wire mesh cup, designed to avoid perturbation of the heat transfer, and suspended from a precision scale, which records the weight. With such a system it is possible to carry out isothermal thermogravimetric analyses of samples with initial weights up to 30 g, in a controlled atmosphere.

The samples employed were pine wood slabs, 5, 7, 9 or 13 mm thick, or pine wood cylinders, 140 mm in length and 15, 20, 22 or 27 mm in diameter.

The procedure is straightforward. The kiln is heated to the desired temperature and a relatively low flow rate of gas (i.e. air, CO2 or N2) passes through the reactor. Then the wood sample, previously dried, is introduced into the hot zone, and the weight and temperature are recorded.

This experimental system allowed us to study the rate of thermal conversion in connection with the gaseous atmosphere, the temperature and the size.

Results and discussion

Influence of the gaseous atmosphere

In a gasifier, pyrolysis is always the first stage, but its rate could depend on the reaction medium, which itself depends on the gasification process.

Figure 4 which gives the weight losses of identical slabs pyrolyses in different atmospheres, leads to some useful qualitative conclusions.

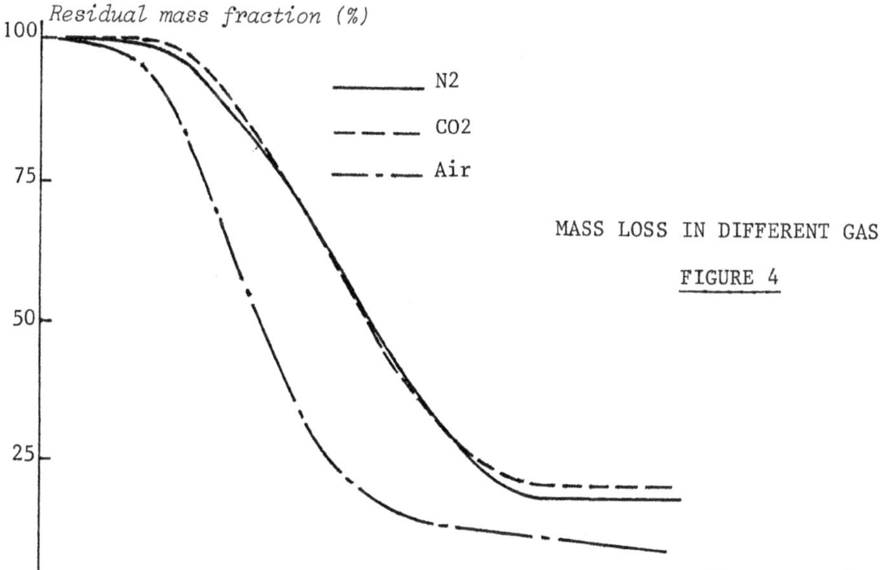

MASS LOSS IN DIFFERENT GASES

FIGURE 4

For a 7 mm slab at 500°C pyrolysis takes place in the same manner whether in nitrogen or carbon dioxide. We find that the volatilization of wood does not depend upon the surrounding gas and that CO2 plays no role in the reaction. It seems that, in an oxidizing atmosphere, an increase in the rate is likely only due to the increase in temperature which takes place with the combustion of the pyrolysis gas at the surface of the solid.

Influence of temperature

For a given size, the pyrolysis time and the weight of the residual solid decrease when the temperature increases.

Influence of size

As shown in Figure 5, the thicker the slab, the more important is the weight of the residual charcoal. This is not surprising because the size affects the rate of heat transfer in the solid. Then the rate of pyrolysis becomes slower, thus improving, the formation of char due to tar decomposition.

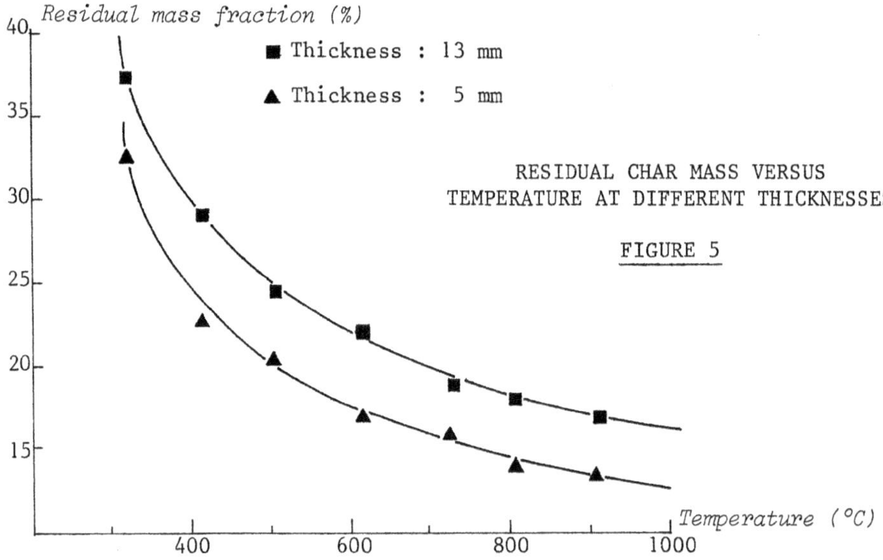

RESIDUAL CHAR MASS VERSUS TEMPERATURE AT DIFFERENT THICKNESSES

FIGURE 5

Moreover, this effect is not so important when the temperature rises.

Figure 6 gives the total time for pyrolysis versus the thickness at different temperatures. It is clear that this time changes linearly with the square of the thickness. The calculated linear correlation coefficients for linear relationships given in Figure 6 are always greater than 0.99. Such numerical results show that, for temperatures higher than 600°C, the rate of pyrolysis is controlled by the rate of heat transfer. Obviously, the slope of the linear relationship is related to α, the thermal diffusivity, and this slope decreases with temperature. Hence, the diffusivity increases when the wood conversion temperature rises.

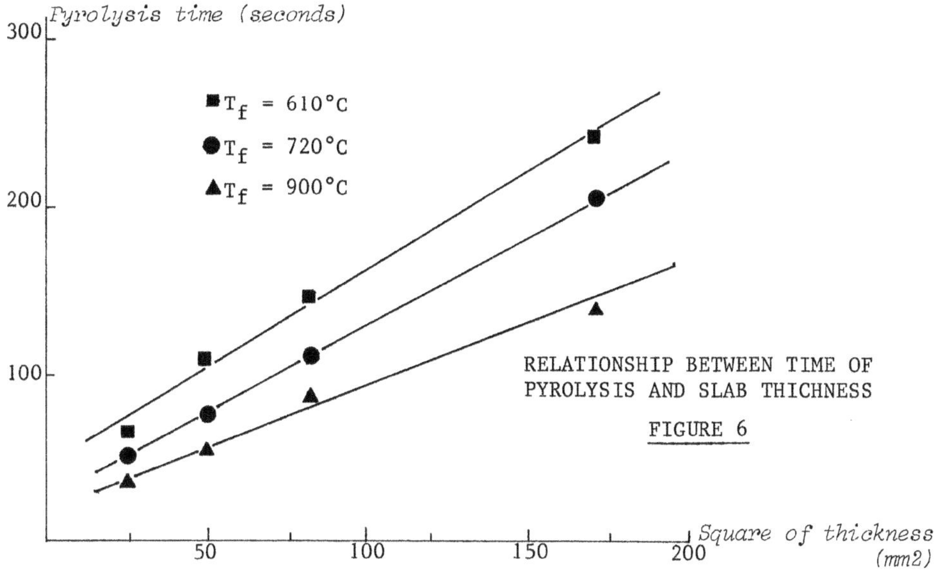

RELATIONSHIP BETWEEN TIME OF PYROLYSIS AND SLAB THICHNESS

FIGURE 6

A simular conclusion has been drawn from experiments with a wood cylinder. Therefore, in order to provide a more quantitative interpretation, we propose a simple model, with only two equations, one for the unsteady heat conduction and another for the first order rate conversion.

4.3.2 Modelling pyrolysis

Our aim is to develop a simple model, with the minimum set of parameters, which enables us to predict, with an acceptable precision, the pyrolysis time of dry wood samples.

The equation for heat transfer

In the experimental procedure, a sample of dry wood, held in a wire mesh cup, is put into a kiln previously heated to a temperature T_f. Heat is transferred to the solid by conduction, convection and radiation. First, we checked that the cup itself had only a negligible influence on the experiment. Next, the effect of the gas flow was studied. Measurements taken with different, but always weak, heating rates showed that pyrolysis was not changed by gas convection in the temperature range 500°C to 900°C. However, because of an external resistance to the heat transfer, the wood surface does not immediately reach the fixed temperature T_f. This response time remains short compared with the time needed for the completion of the pyrolysis (it was equal to 10 s at 840°C and to 15 s at 630°C). Therefore, in the definition of initial conditions, we have assumed that the surface temperature of the wood sample follows a linear time-profile, with a steep slope, to reach the fixed temperature T_f.

The temperature distribution in the wood also depends on the heat of reaction, the value of which, for wood pyrolysis, is still controversial.

Differential thermal analysis (DTA) and differential scanning calorimetry (DSC) results support the contention that the heat of pyrolysis of wood is slightly endothermic, at least where secondary reactions are not important. Heat effects change with the temperature, but remain weak. We have therefore not included the thermicity of the reaction in the heat balance, which is written, for a cylinder as :

$$\frac{\partial T}{\partial t} = \alpha \frac{\partial^2 T}{\partial r^2} + \frac{1}{r} \left(\frac{\partial T}{\partial r}\right)$$

with the following conditions :

Initial condition : $t = 0,\ T(r,0) = T_0$

Boundary conditions : $t < t_i,\ T(R,t) = \frac{t}{ti} (T_f - T_0) + T_0$

$t > t_i,\ T(R,t) = T_f$

This equation is solved by a finite difference procedure assuming cylindrical symmetry.

The main problem is the choice of a numerical value of the thermal diffusivity, which changes significantly during volatilization. For dry wood, Tinney proposed $\alpha = 16 - 10^{-8} m^2 s^{-1}$ and, for charcoal, the α value is much larger.

If we look at Figure 7, which shows the variation of the centre temperature of a heated wood cylinder as a function of time, we can see three regions : the temperature increases sharply, then, between 130°C and 450°C, the curve looks like a classical curve that would be obtained in a common unsteady heat diffusion problem. Therefore, to fit the experimental temperature distribution we choose, after trial-and-error optimizations, three values for the thermal diffusivity :

T < 130°C	$\alpha = 16 \times 10^{-8} m^2 s^{-1}$
170°C < T < 450°C	$\alpha = 9 \times 10^{-8} m^2 s^{-1}$
600°C <	$\alpha = 63 \times 10^{-8} m^2 s^{-1}$

Between these two regions, the parameter was assumed to vary linearly.

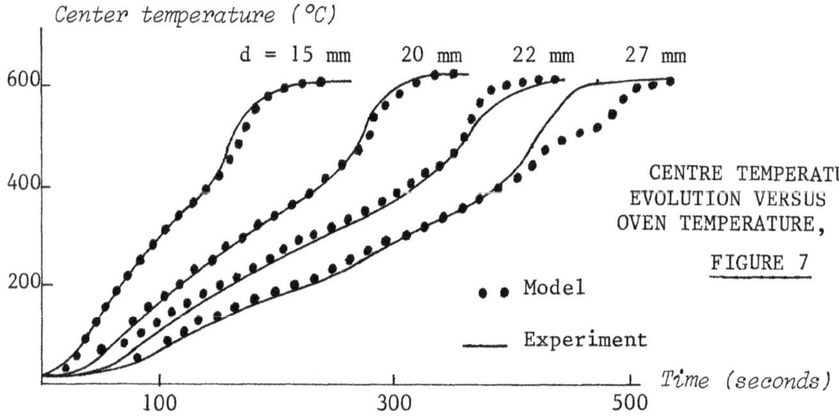

CENTRE TEMPERATURE EVOLUTION VERSUS TIME OVEN TEMPERATURE, 600°C

FIGURE 7

We can predict the temperature distribution inside the wood cylinder, and Figure 7 gives the results for T_f, 630°C and four different diameters. It can be noted that this simple method of calculation gives numerical results which agree well with experimental ones.

The values used for thermal diffusivity need to be discussed. For temperatures below 130°C, the value is that for raw wood. In the range 170°C to 450°C, carbonization occurs and pyrolysis gases escape, leading to a convective outflow which limits heat conduction and therefore the global effect is decrease in the apparent diffusivity. Above 450°C, carbonization has ended and diffusivity reaches the usual value for wood char. The value found by optimization is $\alpha = 64 \times 10^{-8} m^2 s^{-1}$. If we calculate the value by using the specific mass and the thermal conductivity given by Havens, and the heat capacity, C_p' proposed by Widell, we obtain :

$$\alpha = 60 \times 10^{-8} m^2 s^{-1}$$

a result which agrees well with the optimization value.

It is now possible to use the temperature distribution to integrate the first-order kinetic equation.

The chemical kinetics

The most simple way to describe the rate of conversion is to assume the reaction to be of first order : i.e. :

$$\frac{d\rho}{dt} = - k(\rho - \rho_f) \text{ with } k = k_0 e^{-E/RT}$$

But conversion is not homogeneous and density is not constant, neither in time nor in space, and so the experimental system provides only a mean value.

If $\rho(r,t)$ is the density function, its mean value is given by :

$$\bar{\rho}(r,t) = \frac{2}{\rho^2}\int_0^\rho \rho(r,t) r dr$$

with :

$$\rho(r,t) = \rho_0 + \int_0^t \frac{\partial\rho(r,t)}{\partial t} dt$$

Using these equations, the mean value of the density can be calculated as a function of the solids residence time, t, and temperature. But these numerical values depend on the kinetic parameters k_o and E. So it is necessary to proceed to an optimization using the Rosenbrock algorithm.

On the basis of four experimental curves, shown in Figures 8 and 9, the calculation gives the following optimum values :

$$k_o = 10^8 s^{-1} \qquad E = 125 \text{ kJ/mole}$$

These values can be compared with those already published by Tang who found E = 120 kJ/mole for the pyrolysis of pine wood cylinders. Furthermore, the overall model, although it is simple, gives a very good description of the rate of high temperature pyrolysis of a wood sample;

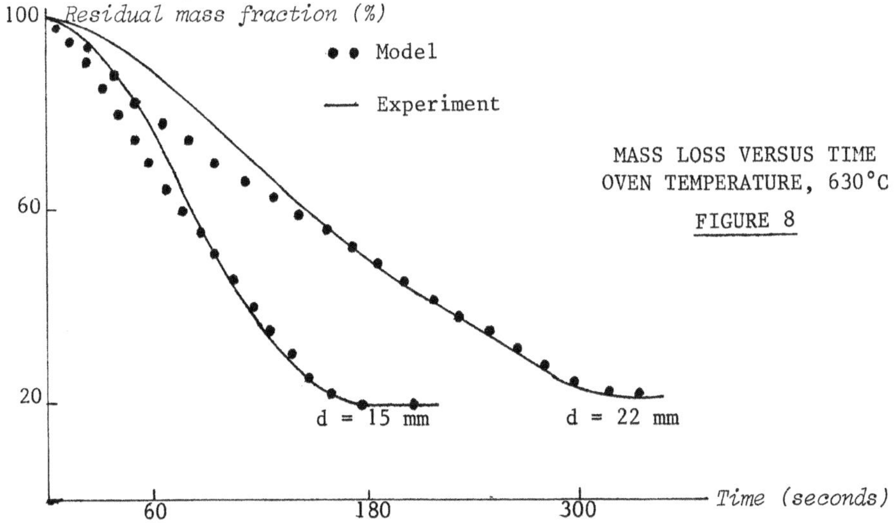

MASS LOSS VERSUS TIME
OVEN TEMPERATURE, 630°C

FIGURE 8

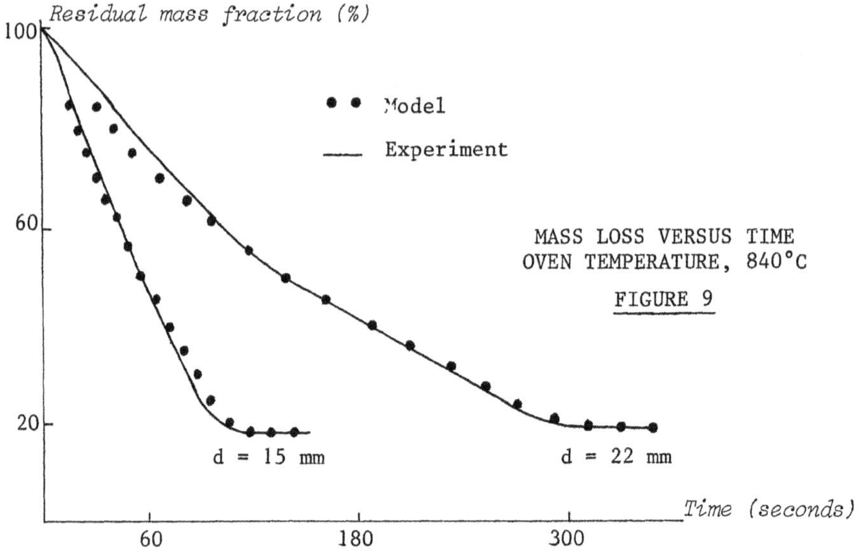

MASS LOSS VERSUS TIME
OVEN TEMPERATURE, 840°C

FIGURE 9

In order to put the results in a more practical form, we have drawn the curves shown in Figure 10, which gives t_p as a function of both size and temperature. These curves could be a very useful tool for scaling up the behaviour of a gasifier.

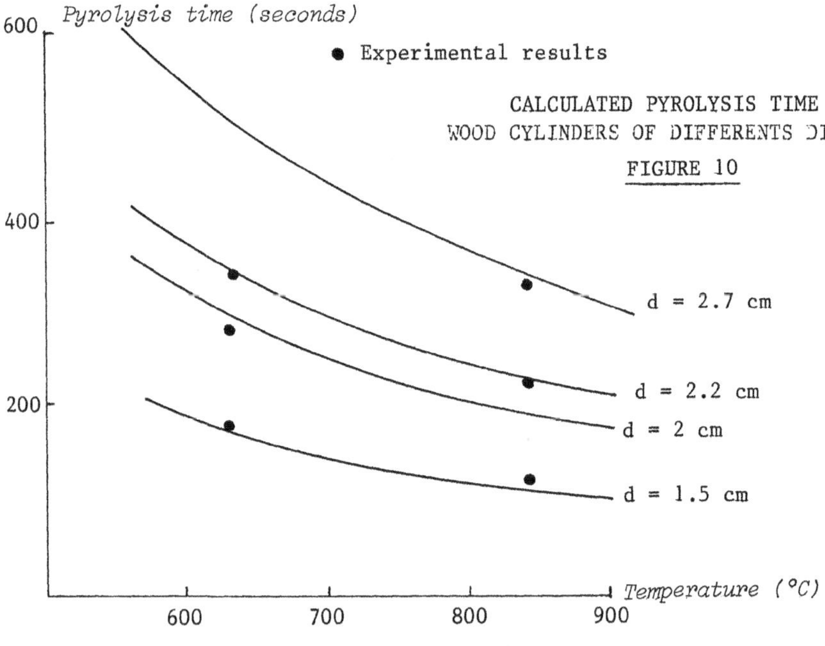

CALCULATED PYROLYSIS TIME OF
WOOD CYLINDERS OF DIFFERENTS DIAMETERS

FIGURE 10

4.3.3 The pyrolysis compounds

The composition of products which are delivered during the pyrolysis is depending upon the type of wood and operating conditions. Two pyrolysis types can be distinguished : the slow and the fast pyrolysis. In the two types, the products obtained will be some gases (CO, CO2, H2, CH4...), a solid residue, and liquid products. In case of a fast pyrolysis -which we can assume for a wood chip falling down into a 700-800°C fluidized bed- the volume of released gases is higher than in case of a slow pyrolysis. On the contrary, liquid products decrease.

To give a picture of the pyrolysis products, we present here some results of flash pyrolysis obtained on Douglas fir and on beech (from Deglise) : wood chemical formula assumed : $CH_{1.44}O_{0.66}$.

$T = 800°C$

$CH_{1.44}O_{0.66} \rightarrow x_1CO + x_2CO2 + \ldots\ldots + x_9$ (residue)

	Douglas fir	Beech
x_1 (CO)	0.340	0.312
x_2 (CO2)	0.031	0.048
x_3 (H2)	0.124	0.112
x_4 (CH4)	0.094	0.091
x_5 (C2H4)	0.043	0.040
x_6 (C2H6)	0.002	0.003
x_7 (C2H2)	0.006	0.005
x_8 (H2O)	0.197	0.165
x_9 (residue)	0.425	0.451
	$[CH_{0.52}O_{0.12}]$	$[CH_{0.730}O_{0.19}]$

The analyses of these results show that at 800°C :

	Douglas fir	Beech
% mass solid	25.6	29.6
% mass gas	74.4	70.4
% volume $\frac{\text{Cn Hm *}}{\text{total gas}}$	17.3	17.9
% volume $\frac{\text{Cn Hm}}{\text{dry gas}}$	22.3	22.7

* Cn Hm = CH4 + C2H6 + C2H4 + C2H2

4.4 Kinetics of the decomposition of methane

The important quantities of light hydrocarbons, particularly methane, released during the high temperature pyrolysis of wood lead us to the question of their decomposition. We studied the case of methane, which prevails due to its high content at the outlet of the first stage of our process.

In fact, the behaviour of wood chips at high temperature lead to the conclusion that methane formation (and more generally hydrocarbons) is unavoidable, and that we have to enquire about its decomposition.

Methane can react with steam by the following reaction :

$CH4 + H2O \rightleftarrows CO + 3H2$

There is not much information in the literature about the homogeneous decomposition of methane. In one paper, Zahradnik and Grace have studied the kinetics of this reaction. They found the following expression :

$$\frac{d\,[CH4]}{[CH4]} = - ko \exp.\left[-\frac{E}{RT}\right] dt$$

with

$ko = 758.47\ s^{-1}$

$r = 3.5766$ cal/mole °R

This leads, for a given value of temperature, to the following expression

$$\frac{[CH4]\ \text{residual}}{[CH4]\ \text{origin}} = \exp\left[- 758.47.t.\exp\left(-\frac{16644}{T}\right)\right]$$

with t in seconds.

That expression can be applied when steam is in excess.
The following Figure 11 shows the residual CH4 as a function of time, at different temperatures.

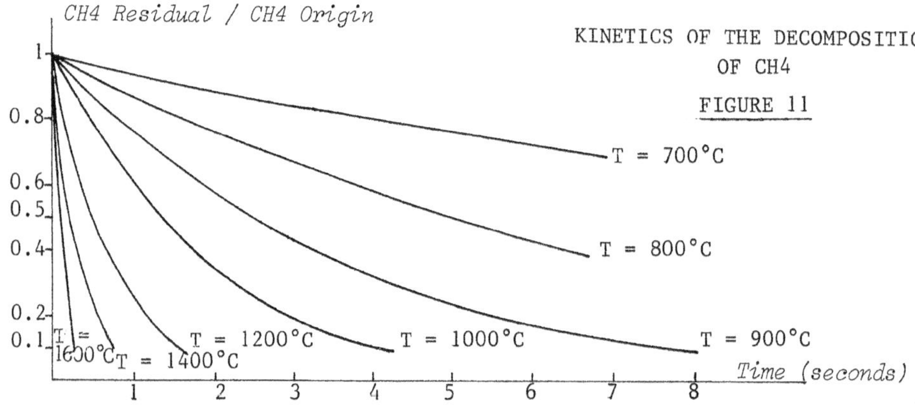

KINETICS OF THE DECOMPOSITION OF CH4

FIGURE 11

5 THERMODYNAMICS OF GASIFICATION REACTIONS

Independently of any kinetics, we checked the thermodynamic equilibriums involved in the wood gasification process. Results hereafter are assumed to be at the equilibrium, and they have been calculated when varying several parameters :

- pressure = 1, 10, 20 bars
- wood moisture = 17, 33, 52 % (on wet wood)

5.1 Assumptions

The general equation of wood conversion is :

$$C_6H_9O_4 + aO_2 + bH_2O \rightarrow xCO + yCO_2 + zH_2 + tH_2O + uCH_4 + vC_s$$

a, b : given
C_s : solid carbon
x, y, z, t, u, v : to be checked

Carbon (as graphite) is the only solid species considered. Information from the literature shows that the only products which are thermodynamically present in significant quantities ($> 10^{-4}$ mole %) in the considered conditions of gasification are C_s, H_2, H_2O, CO, CO_2 and CH_4.

It is interesting to notice that no other hydrocarbon than CH_4 is thermodynamically stable in gasification conditions. Acetylene, ethylene and other hydrocarbons are the results, when they are present, of a "non-equilibrium".

Material balance

Carbon : $6 = x + y + u + v$
Hydrogen : $9 + 2b = 2z + 2t + 4u$
Oxygen : $4 + 2a + b = x + 2y + t$

Equations of equilibrium

$CO + H_2O \rightleftarrows CO_2 + H_2$	$k_1 = K_3/K_2$
$C_S + CO_2 \rightleftarrows 2CO$	K_2
$C_s + H_2O \rightleftarrows CO + H_2O$	k_3
$CO + 3H_2 \rightleftarrows CH_4 + H_2O$	$k_4 = K_5/k_3$
$C_s + 2H_2 \rightleftarrows CH_4$	k_5

Thermal balance

Heat coming in = heat coming out + thermal losses

5.2 Resolution of the system

The method consists to choose a value for oxygen and to make interations on temperature.

The system is considered being solved when the thermal balance is satisfied.

5.3 Results

They are represented in figures 12 to 20, and give the gas composition and the oxygen required as a function of the temperature of reactions.

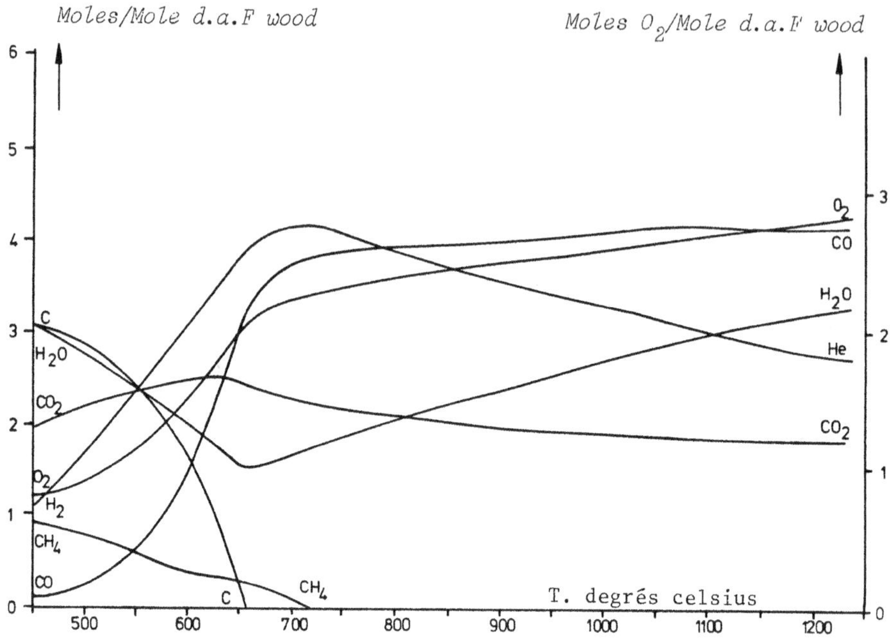

FIGURE 12

Pressure = 1 bar
Wood moisture = 17%/raw

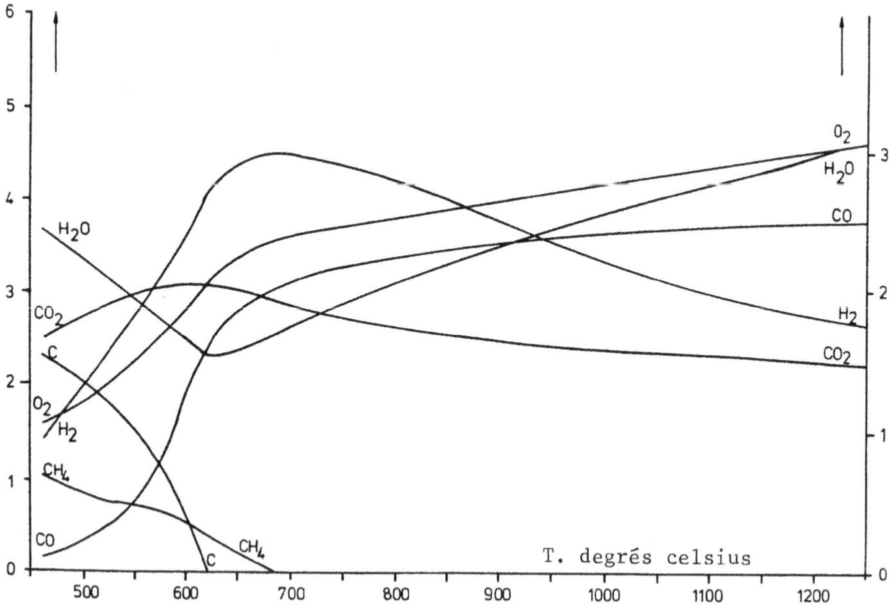

FIGURE 13

Pressure = 1 bar
Wood moisture = 33%/raw

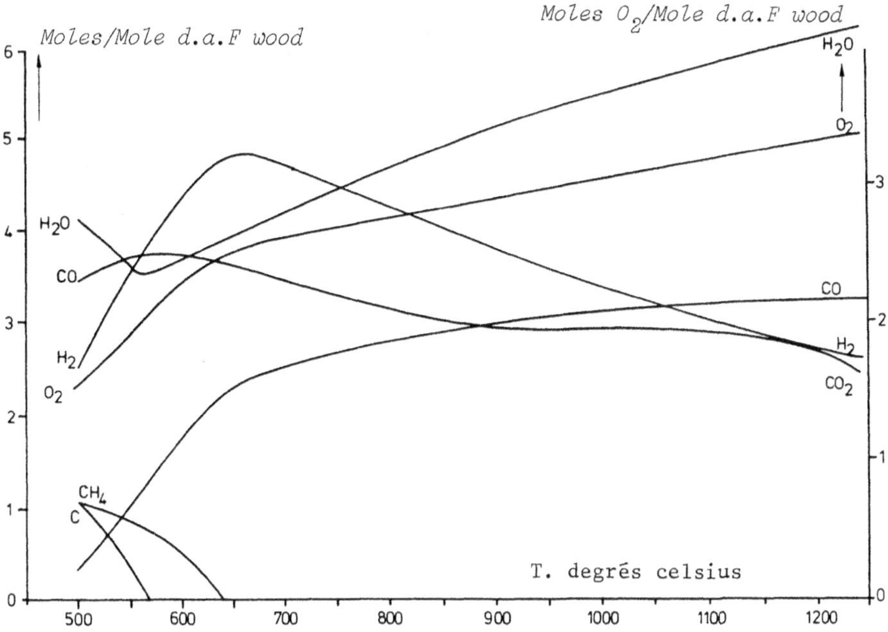

FIGURE 14

Pressure = 1 bar
Wood moisture = 52%/raw

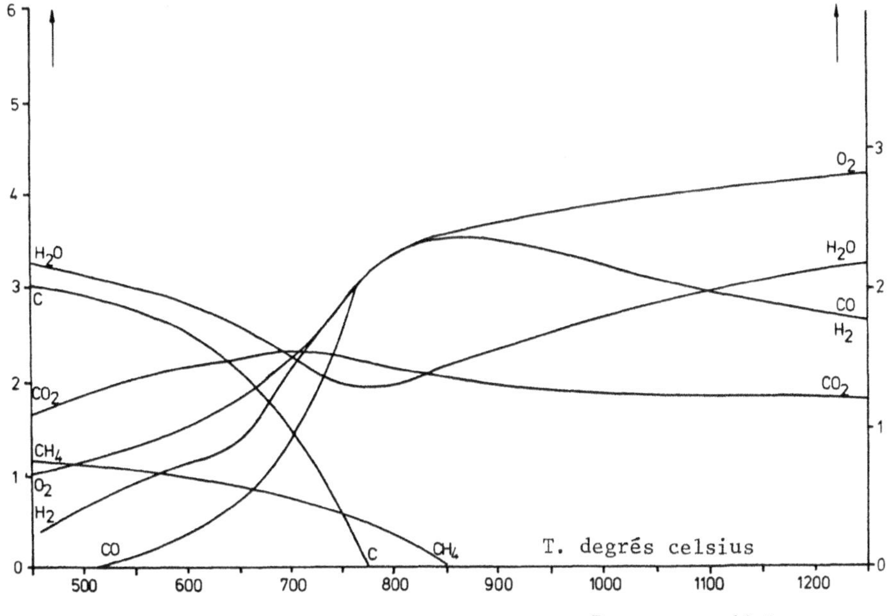

FIGURE 15

Pressure = 10 bar
Wood moisture = 17%/raw

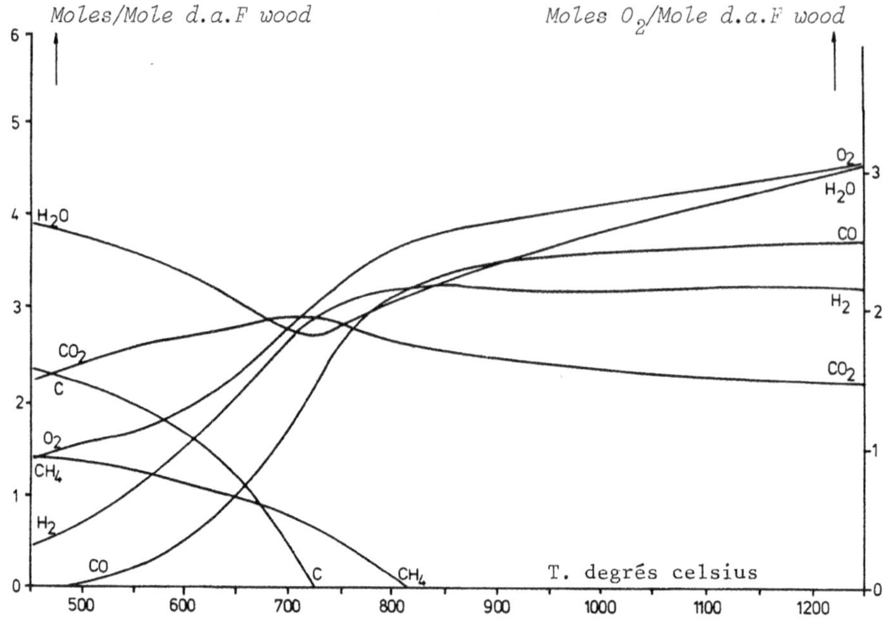

FIGURE 16

Pressure = 10 bar
Wood moisture = 33%/raw

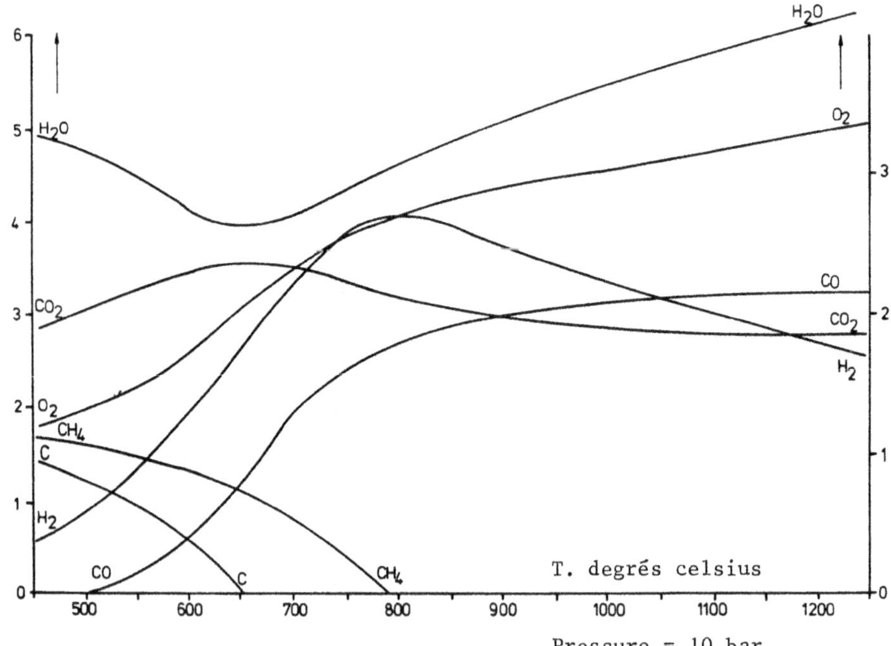

FIGURE 17

Pressure = 10 bar
Wood moisture = 52%/raw

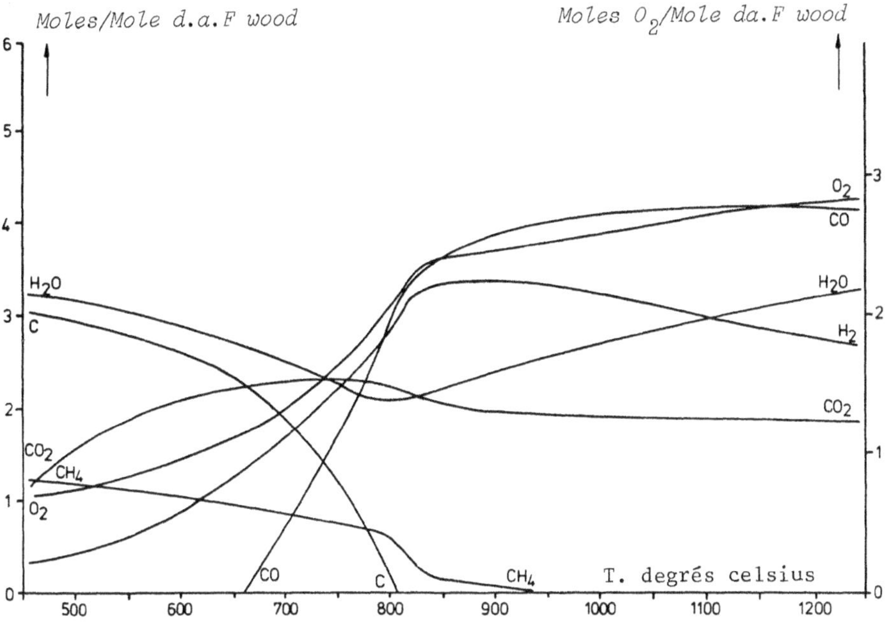

FIGURE 18

Pressure = 20 bar
Wood moisture = 17%/raw

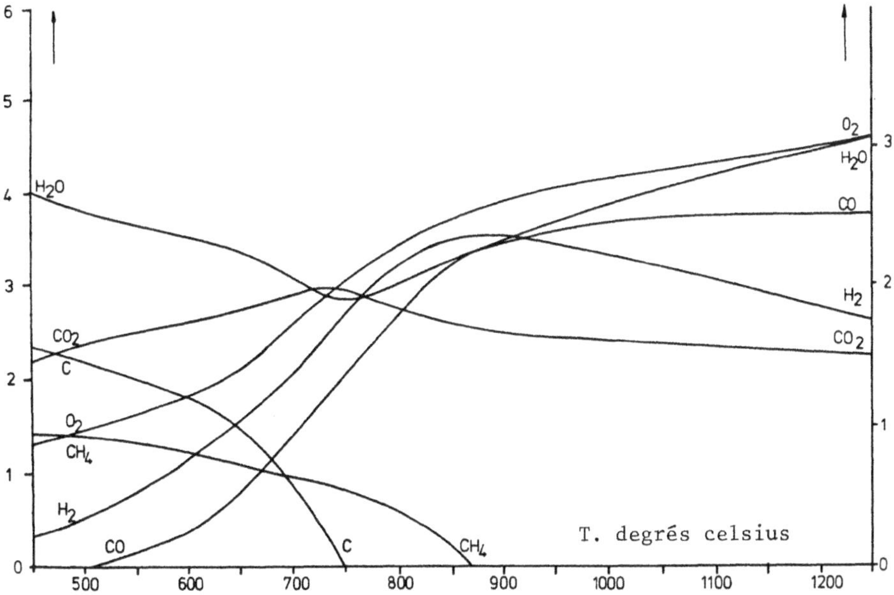

FIGURE 19

Pressure = 20 bar
Wood moisture = 33%/raw

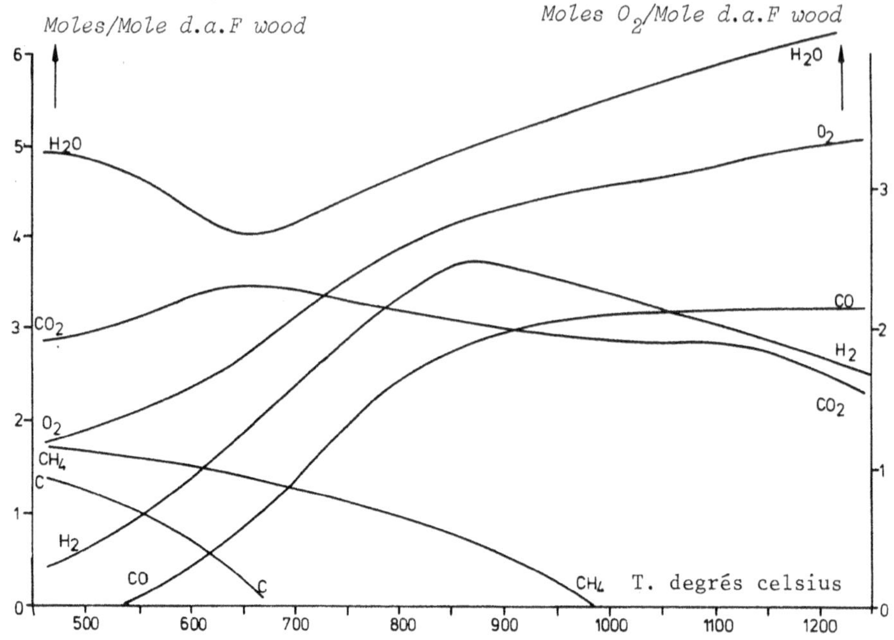

FIGURE 20

Pressure = 20 bar
Wood moisture = 52%/raw

Oxygen consumption

Stoechiometric combustion of wood corresponds to the reaction :

$C_6H_9O_4 + 6.25\ O_2 \rightarrow 6\ CO_2 + 4.5\ H_2O$

Ideal gasification of wood, from a thermodynamical point of view, would proceed according to :

$C_6H_9O_4 + O_2 \rightarrow 5\ CO + 4.5\ H_2$

On the figures 12 to 20 is represented, among other curves, the oxygen demand as a function of temperature. The increase in oxygen consumption can roughly be divided into two parts : the first one were temperature rises moderately with oxygen and the second one where it rises very rapidly. The transition point corresponds, in fact, with the complete disappearance of solid carbon.

At 1 200°C, the "thermodynamical" oxygen demand is 0.6 kg/kg dry wood.

Influence of wood moisture

The oxygen demand increases with the water content of wood. The oxygen consumption can be increased by 25 % when wood moisture jumps from 15 to 50 %.

Influence of pressure

For the same gasification temperature and wood moisture, the increase of pressure leads to a very low decrease of oxygen.

In fact, the gas composition is only slightly influenced by pressure.

6 EXPERIMENTS WITH THE ATMOSPHERIC PILOT PLANT

As it has been mentioned earlier (see the paragraph 1.1 "Basic research on the pilot"), the atmospheric pilot of Le Creusot was modified during the past four years of experiments. The main steps in the evolution of the pilot are :

End of 1980 : Construction of a single fluidized bed gasifier, 400 mm internal diameter, fed with wood chips and pure oxygen. Dry-wood capacity = 2.5 - 5.0 t/day.

Mid 1982 : Extension of the single fluidized bed gasifier by a secondary reformer (for partial oxidation of the raw gas leaving the gasifier), 500 mm internal diameter.

End of 1983 : Modification of the atmospheric pilot to rise its capacity up to 10 t/day, and to limit the nitrogen content of syngas.

6.1 Overview of the whole experimental program

During the first year of testings on the single fluidized bed gasifier (0.4 m I.D), the wood feed-rate varied between 100 and 200 kg/h and its moisture between 20 and 40 % (on a wet basis). Pure oxygen was injected at the bottom of the fluidized bed made of silica sand. No steam was injected. Experiments were made at different temperatures, between 700 and 850°C. Due to the presence of hot spots in the fluidized bed, some sintering problems appeared in the range 800-850°C, even for short runs.

To prevent sintering, several improvements were made :

1 - Change in type of inert material. The silica sand was then replaced by a finely divided fire-clay material.

2 - Modification of the injection of oxygen, which was replaced by an injection of a mixture of steam + oxygen, in order to limit the formation of hot spots, and also the significantly increase the fluidizing gas velocity.

3 - From now on, experiments were conducted below 800°C, and in those modified operating conditions, a satisfactory behaviour of the fluid bed was observed.

From the experiments with the single fluidized bed only, the main analytical results were :

- The CO and CO2 content of the gas leaving the gasifier is very dependent on wood moisture. The relative contribution of exothermal reactions increases with water content of gas in order to maintain the adiabatic working conditions. In order to be more representative for the industrial application, it was decided to operate, when possible, on wood of 25 % maximum moisture.
- The H2 content of the dry gas leaving the single gasifier most of time was in the range 15-20 %, whatever the moisture content.
- The most relevant result concerning the gas leaving that primary stage of the process is that the hydrocarbon content is high : CH_4 = 10-12 %, C_2H_4 = 2-3 %... As shown in the previous chapter, that high C_nH_m content of the raw gas is the result of the high volatile matter content of wood which delivers about 80 % of its mass during pyrolysis.
- Another significant result that we obtained at the outlet of the gasifier is that, due to the rather low temperature at which the reactor is operated, significant amounts of oil + tars + unburnt carbon are produced.

As the fluidized bed temperature can not be maintained above 800°C because of ash sintering problems, it was necessary to conceive a second step of treatment in order to make our process suitable for syngas production.

Then, a secondary reformer (which is, properly speaking, a partial oxidation reactor) was added to the fluidized bed gasifier (Mid 1982).

Secondary oxygen was injected in the gas entering that refractory lined reactor, leading to a partial combustion and then to an increase of the temperature up to 1 300-1 450°C.

With that complete atmospheric pilot plant (fluidized bed = 400 mm I.D– secondary reformer = 500 mm I.D), three long duration runs, 100 hours continuous for each one, where performed in November 1982, December 1982, March 1983.

These long duration runs allowed :

- To put to the test the overall pilot plant on long periods.
- To test the enrichment of the fluid bed with wood ashes and to prevent sintering conditions. We here point out that such a problem is one of the most critical in oxygen gasification.
- To optimize the working conditions of the secondary reformer, on the following items :
  . mode of injection of the gas leaving the gasifier
  . mode of injection of oxygen
  . temperature
  . selection of refractories
- To verify that the assumptions used for the pressurized pilot plant of Clamecy (60 TPD dry wood) were correct.
- To demonstrate the influence of some parameters, such as the wood moisture, on the gas characteristics, and then to validate the necessity of a preliminary wood drying down to about 15 %.
- Finally, to obtain experimental results on the quality of quench water from the syngas.

* Presentation of the three long duration runs

Test n° 70 : November, 22-26th, 1982
Hardwood, 20 % moisture, granulometry 10–20 mm
Fluidized bed : silica sand
Gasification medium : oxygen + steam
Fluidized bed temperature : 700°C ± 20°C
Secondary reformer with an upward flow of gases
Tangential inlets of gas and oxygen at the bottom of the reformer
Maximum temperature : 1 300°C

Conclusions of the test n° 70
Gas characteristics satisfactory
Local damages on refractories
Slow enrichment of the fluidized bed with wood ashes, without any bed agglomeration after 100 hours

Test n° 72 : December, 13–17th, 1982
Hardwood, coming from the region of Clamecy, 40 % moisture, granulometry < 50 mm
Fluidized bed : silica sand
Gasification medium : oxygen + steam
Fluidized bed temperature : 700°C ± 20°C during 75 hours
750°C ± 20°C during 20 hours
Secondary reformer with an upward flow of gases
Tangential inlets of gas and oxygen at the bottom of the secondary reformer
Maximum temperature : 1 200°C

Conclusions of the test n° 72
Bad gas characteristics, due to the high wood moisture → very bad potential methanol yield
Confirmation of local damages on refractories, which oblige us to limit the operating temperature at 1 200°C.
Slow enrichment of the fluidized bed with wood ashes

Test n° 75 : March 28th to April 1st 1983
Hardwood, 20 % moisture, granulometry 10–20 mm
Fluidized bed temperature : 700°C ± 20°C

Downward flow of gases in the secondary reformer (the same as in the pilot plant of Clamecy)
Tangential inlet of gas to be reformed
Oxygen inlet at the top of the secondary reformer
High quality refractory lining
Maximum temperature : 1 350°C

Conclusions of the test n° 75
Gas characteristics satisfactory
Refractories not damaged
The following table allows to make a comparison between the expected theoretical results on the pressurized pilot plant of Clamecy and the experimental results obtained during the third long duration run n° 75
The operating conditions were as follow :
- wet wood feed-rate : 109.7 kg/h
- wood moisture : 20 % (wet basis)
- gasification temperature : 700°C
- operating time under stablilized conditions : 95 hours

Remark : in this table, each of the analysis numbered 900 to 2 300 is the result of an average calculated each time on 7 to 9 gas analyses.

TABLE OF RESULTS (LONG DURATION RUN N° 75)						Case of Clamecy (assumption)
Date		29/03/83	30/03/83	31/03/83	01/04/83	
Analysis n°		900	1 500	1 700	2 300	
Dry wood feed-rate kg/h		87.80	87.80	87.80	87.80	87.80
Reformer 02 flow-rate Nm3/h		17.50	13.50	13.00	17.50	
Fluidized bed 02 flow-rate Nm3/h		14.50	15.50	15.00	16.50	
kg 02/kg dry wood		0.52	0.47	0.46	0.55	0.60
Equivalence ratio *		0.38	0.34	0.33	0.39	0.43
Steam 200°C/dry wood		0.23	0.23	0.23	0.23	0.40
Wood moisture		20.00	20.00	20.00	20.00	15.00
Balance C - H - O oultel/inlet %	C	100.00	95.00	96.00	96.00	100.00
	H	85.00	89.00	87.00	95.00	100.00
	O	101.00	101.00	101.00	100.00	100.00
Dry gas flow-rate (corrected) (CO + CO2 + H2 + CH4) Nm3/h		121.80	115.20	117.50	121.10	117.70
% corrected/dry gas	CO	34.00	30.60	31.40	35.00	38.80
	CO2	31.50	32.90	32.10	28.20	27.40
	H2	31.50	31.90	32.90	33.60	30.90
	CH4	1.60	3.90	3.20	1.50	3.00
Corrected thermal yield % LHV gas / LHV wood *		63.00	64.00	65.00	65.00	69.00
Thermal balance (without thermal losses) (%) *		81.00	79.00	79.00	81.00	97.00
Tons wood gasified per ton of methanol		2.31	2.56	2.44	2.22	2.25

* Notes :

1 - Equivalence ratio = $\dfrac{\text{Oxygen really consumed}}{\text{Oxygen that should be consumed in the stoechiometric combustion of wood}}$

2 - "Corrected thermal yield" means that it takes into account the ratio between the L.H.V of the corrected dry gas flow-rate and the L.H.V of wood.

3 - Thermal balance without thermal losses is :

$$\frac{\text{output heat}}{\text{input heat}} = \frac{\text{(heating value + specific heat at the outlet temperature) products}}{\text{(heating value + specific heat) wood input}}$$

The analyses 900 and 2 300, performed with an oxygen flow rate comparable to bhe basis case for the design of the pressurized unit of Clamecy, point out that the "potential methanol" is very close to the theoretical value.

Under the same conditions, we observe that the CH_4 content is decreased down to 1.5 %, all higher hydrocarbons (C_2H_4, C_2H_6, C_3H_8, oil and tars) are traces only, not detectable by the analysis. Moreover, the carbon balance proves that no soot is formed, which is confirmed by the quench water analysis (less than 2 % unburnt carbon).

The results of analyses 900 and 2 300 (reformer temperature > 1 300°C) when compared to those of analysis 1 500 and 1 700 (reformer temperature ≃ 1 150°C) indicate that it is necessary, in order to reach a CH_4 content lower than 2 %, to operate over 1 300°C.

Conclusions after these three long duration runs

The optimization phase of the P.D.U. of Le Creusot is well advanced. Main operating conditions are a quite moderate temperature of the fluidized bed (700-800°C) and a high temperature in the secondary reformer (1 300 - 1 400°C).

In the fluidized bed, steam acts as a fluidization medium and participates as a gasification reagent. The methanol yield just slightly decreases with an increase in steam rate.

In the secondary reformer, the mode of injection for both gas and oxygen was established. The methane content of the syngas is sufficiently low to methanol application. Moreover, the unburnt carbon production is low. The oil and tars production is negligible.

The potential methanol yield from gas produced by the P.D.U. agrees with the assumptions made for the design of the pressurized unit of Clamecy.

6.2 Bonus test run

As mentioned in chapter 2 ("definition of the program"), a contractual guarantee run had to be performed for the C.E.C. stipulating the following main conditions :

- Syngas samples will be taken at regular intervals during a 24 hours period of gasification. These samples will be taken in the presence of experts of the C.E.C.
- Samples of syngas should contain :

. at least 70 % volume of hydrogen + carbon monoxide

. less than 3 % volume of inerts (nitrogen + methane + uncondensable hydrocarbons)

. the installation should gasify 400 kg/h of wood, 20 % moisture

To meet these requirements, we modified the atmospheric pilot so that its capacity rose up to 400 kg/h of wood and the inert content of the dry gas decreased below 3 %. These modifications are described in chapter 3 (description of the atmospheric pilot plant of Le Creusot).

6.2.1 Introduction

Initially scheduled to begin on March, 6th, the 24 hours guarantee run had to be delayed because of a plug of solids at the outlet of the water scrubber, due to solids build up after the preliminary testings of February.

Then, the atmospheric pilot was preheated with natural gas during the night between 6th and 7th of March 1984. Wood gasification then started up, at the pre-determined conditions of fluid bed temperature and secondary reformer, at 7 h 30 a.m. on March 7th.

The nominal wood feed-rate was reached at 9 h 34 a.m.

Our wood storage for that 24 hours test was divided into two bins ; the first one was stored in the laboratory itself, and its moisture was quite regular around 15 % ; the other wood-bin was stored outside and its moisture was very heterogeneous, since it varied between 10 and 37 %. The consequence of such a variation was that the second part of the test run, after 1 h a.m. on March the 8th was more delicate.

We just observed one incident during the test run. At 11 h 15 p.m. on March the 7th, the rotary valve of the wood feeding system blocked, leading to an automatic shut-off of the oxygen and wood feeding. The installation was then placed in the "hot stand-by" state during the 2 hours of intervention and then started again with wood and oxygen feeding.

The test run was then continued up to March 8th at 1 h p.m.

During the overall test run, 88 gas chromatographic analyses were performed.

6.2.2 - Average results

- Start-up of wood feeding : March, 7th, 7 h 30 a.m.
- End of wood feeding : March, 8th, 1 h 04 p.m.

Incident : shut-off of wood feeding during 1 h 50 mm (between March, 7th at 11 h 14 p.m. and March, 8th at 1 h 05 a.m.), caused by the blocking of the rotary valve

- Total time of the test run in the nominal conditions : 24 h 40 mn
- Wood moisture :

. March, 7th (storage inside the laboratory : 3 samples → 15.1 % - 14.8 % - 12.8 %

. March, 8th (storage outside the laboratory) : 5 samples → 20.6 % (2 h 45 a.m.) - 20.9 % (6 h a.m.) - 35.5 % (7 h 30 a.m.) - 37.3 % (8 h 30 a.m.) - 25.4 % (10 h 30 a.m.)

- Number of gas chromatrographic analyses : 88
- Average composition of dry gas based on these 88 analyses :

(% vol./dry gas)

	Raw result	Corrected result to 100 %	Correction N2, Ar
O2	38.89	39.29	41.10
H2	30.44	30.76	32.17
CO2	24.71	24.97	26.12
CH4	0.57	0.58	0.61
Ar	0.53	0.53	-
N2	3.83	3.87	-
TOTAL	98.94	100.00	100.00
CO + H2	69.33	70.05	73.27

- To give an idea of the dispersion of the gas composition around that average result, we present, from these 88 analyses, the best and the worse one below (% vol./dry gas)

Best analysis : n° 16 March 7th, 1 h 45 p.m.	Worse analysis : n° 78 March 8th, 8 h 30 a.m.
CO = 43.00 H2 = 32.60 CO2 = 22.05 CH4 = 0.87 Ar = 0.16 N2 = 0.83 TOTAL = 99.51 CO + H2 = 75.6 % volume CO + H2 (corrected) = 76.7 % volume	CO = 35.20 H2 = 28.08 CO2 = 28.18 CH4 = 0.39 Ar = 0.25 N2 = 6.36 TOTAL = 98.46 CO + H2 = 63.28 % volume CO + H2 (corrected) = 68.9 % volume
Wood moisture = 14 %	Wood moisture = 37 %

Note : the results of these two analyses n° 16 and n° 78 confirm the interest of a preliminary drying of wood at about 15 % moisture.

- In the following table, we present a comparison between the gas chromatographic analysis of CREUSOT-ENERGIE and the mass spectrometer analysis performed by Gaz de France (samples taken on March 7th, in presence of an expert of the C.E.C.).

Time	11 h 35 a.m.		2 h 05 p.m.	
	CREUSOT-ENERGIE (chromato.)	GAZ DE FRANCE (spectro.)	CREUSOT-ENERGIE (chromato.)	GAZ DE FRANCE (spectro.)
Sample n°	7	1.1	17	5.1
% vol./dry gas				
CO	40.57	41.6	41.30	41.6
H2	30.01	30.5	30.12	29.5
CO2	23.55	23.4	22.27	23.5
CH4	0.30	0.4	0.63	0.4
O2	-	0.1	-	0.1
Ar	0.24	0.2	0.23	0.2
N2	4.30	3.8	4.78	4.7
TOTAL	98.99	100.0	99.35	100.0
CO + H2	70.58	72.1	71.42	71.1

6.2.3 Material and energy balance

During the 13 first hours of the test run where we worked with homogeneous wood, the moisture was about 14 %.

During the 12 last hours of the test run, the wood moisture was very heterogeneous (fluctuating in the range 20 % - 37.3 %). Results were very much affected by this and it was quite difficult to check with a good precision the wood moisture corresponding to each gas analysis.

Below, we present the material and energy balance which is based on the 13 first hours of the test run, where 43 gas analyses were delivered.

6.2.3.1 Wood feed-rate

Measured over 64 minutes period, on March the 7th, between 6 h 38 p.m. and 7 h 42 p.m.

Raw wood-rate : 407 kg/h

$$\text{Average moisture} : \frac{15.1 + 14.5 + 12.8}{3} = 14{,}1\ \%\ \text{on wet wood}$$

6.2.3.2 O2 flow-rate

Average over 13 hours

To the fluid bed : 48 Nm3/h

To the secondary reformer : 92.8 Nm3/h

6.2.3.3 Steam flow-rate

Averaged over the 13 hours of the test run, at 30 kg/h.

That steam fed to the gasifier, mixed with oxygen, was slightly superheated at about 200°C.

6.2.3.4 Dry gas production

The injection into the gasifier of an argon tracer, at a controlled flow-rate of 11 Nm3/h, allows to check the dry gas flow-rate produced in the pilot plant. That injection is performed during analyses number 23, 24 and 25.

% volume Ar when injected : 2.43 %/dry gas

% volume Ar when "not" injected : 0.27 %/dry gas

(based on the average on 5 analyses before and 5 analyses after argon injection)

$$\text{Dry gas flow-rate} = 11 \times \frac{100}{2.43 - 0.27} = 509\ \text{Nm3/h}$$

6.2.3.5 Average gas composition (analyses 23, 24, 25)

	Raw %/dry gas	Corrected 100 %	Flow-rate Nm3/h	C kg/h	H kg/h	O kg/h
CO	40.00	40.58	206.55	110.65		147.54
CO2	21.71	22.02	112.08	60.04		160.11
H2	29.52	29.94	152.39		13.61	
CH4	0.58	0.59	3.00	1.61	0.54	
Ar	2.43	2.47	12.57			
N2	4.34	4.40	22.40			
TOTAL	98.58	100.00	508.99	172.30	14.15	307.65

6.2.3.6 O_2 flow-rate entering with N_2

A small quantity of "uncontrolled" air enters the gasifier with wood. The above results give N_2 flow-rate = 22.4 Nm3/h, leading to a "parasit" oxygen flow-rate of 5.95 Nm3/h.

6.2.3.7 C - H - O balances

$$C_{in} = 407 \times \frac{100 - 14.1}{100} \times \frac{72}{145} = 173.6 \text{ kg/h (dry wood} = C_6H_9O_4)$$

C_{out} = 172,3 kg/h (see § 5.2.3.1)

⟹ **Carbon conversion = 99,3 %**

$$O_{in} = \underbrace{\left[407 \times \frac{100 - 14.1}{100} \times \frac{64}{145}\right]}_{\text{Dry wood}} + \underbrace{\left[407 \times \frac{14.1}{100} \times \frac{16}{18}\right]}_{\text{Wood moisture}} + \underbrace{\left[(48 + 92,8 + 4,95) \times \frac{32}{22.4}\right]}_{O_2 \text{ fluid bed} + O_2 \text{ secondary ref.} + \text{"parasit" } O_2} + \underbrace{30 \times \frac{16}{18}}_{\text{steam}}$$

$= \underline{436.2 \text{ kg/h}}$

As we don't measure the water content of syngas at the outlet of the secondary reformer, we here make the assumption that oxygen out = oxygen in. Then, we obtain by deduction the water content of the syngas leaving the secondary reformer.

O_{out} = 436.2 kg/h

O_{out} (dry gas) = 307.65 kg/h (see § 6.2.3.5)

⟹ O_{out} (steam) = 436.2 - 307.65 = 128.55 kg/h

⟹ **Steam in the syngas = 180 Nm3/h** = 144.6 kg/h

$$H_{in} = \underbrace{\left[407 \times \frac{100 - 14.1}{100} \times \frac{9}{145}\right]}_{\text{dry wood}} + \underbrace{\left[407 \times \frac{14.1}{100} \times \frac{2}{18}\right]}_{\text{wood moisture}} + \underbrace{\left[30 \times \frac{2}{18}\right]}_{\text{steam}} = 31.41 \text{ kg/h}$$

$$H_{out} = \underbrace{\left[14.15\right]}_{\text{dry gas (see § 6.2.3.5)}} + \left[144.6 \times \frac{2}{18}\right] = 30.22 \text{ kg/h}$$

⟹ **Hydrogen balance = 96.2 %**

6.2.3.8 O_2 consumption (average on the 13 hours)

O_2 = 48 + 92.8 = 140.8 Nm3/h = 201 kg/h

$$\text{Dry wood} = 407 \times \frac{100 - 14.1}{100} = 350 \text{ kg/h}$$

⟹ **Pure O_2 = 0.57 kg/kg dry wood**

6.2.3.9 Equilibrium

The chemical equilibrium which we consider here is the water gas shift

$$CO_2 + H_2 \underset{kp}{\rightleftharpoons} CO + H_2O$$

$$kp = \frac{CO \quad H2O}{H2O \quad H2} = \frac{206.55 \times 180}{112.08 \times 152.39} = 2.18$$

The value of kp is corresponding to a theoretical temperature of 1100°C. In fact, we observed that the syngas temperature at the outlet of the secondary reformer, before its quench, is about 1200°C.

6.2.3.10 Potential methanol

CO = 206.55 Nm3/h
H2 = 152.39 Nm3/h

$$\text{Potential methanol} = \frac{206.55 + 152.39}{3} = 119.6 \text{ Nm3/h} = 170.9 \text{ kg/h}$$

Potential methanol = 0.49 kg/kg dry wood

6.2.3.11 Thermal balance

Dry wood (L.H.V.) = 4 386 x 350 = 1.53 x 10^6 kcal/h
Syngas (L.H.V.) = (206.55 x 3 060) + (152.39 x 2 610) + (3 x 8 174)
= 1.06 x 10^6 kcal/h

The thermal yield based on L.H.V. is 69 %.

The overall thermal balance, based on a temperature of 1 210°C for the syngas leaving the secondary reformer, and without taking into account the thermal losses of the pilot gives a balance of 93 %.

6.2.3.12 Result of the 43 gas analyses

For these 13 hours performed with 14 % moisture wood, the average gas composition is :
(% vol./dry gas)

	Raw result	Corrected result to 100 %	Corrected Ar, N2
CO	40.47	41.09	43.05
H2	30.44	30.90	32.37
CO2	22.48	22.82	23.91
CH4	0.63	0.64	0.67
Ar	0.43	0.44	-
N2	4.05	4.11	-
TOTAL	98.50	100.00	100.00
CO + H2	70.91	71.99	75.42

Average fluid bed temperature : 740°C
Average temperature gas leaving the gasifier : 680°C
Average temperature gas entering the reformer : 670°C
Range of temperature in the secondary reformer : 1 210 to 1 420°C

In the following table, we compare the experimental results obtained on the atmospheric pilot plant with the results expected with the pressurized unit of Clamecy. This comparison can be considered as the most representative up to now, because the test run has been performed with a wood of about 15 % moisture.

	Atmospheric pilot (Le Creusot)	Pressurized pilot (Clamecy)
Status	PROVEN	ESTIMATED
Wood moisture (%)	15	15
Fluid bed T (°C)	700 - 750	700 - 750
Secondary reformer T (°C)	1 210 - 1 450	1 300
O_2 total/wood (kg/kg dry)	0.57	0.60
Steam/wood (kg/kg dry)	0.08	0.40
Carbon conversion (%)	99.3	100
Dry gas production (Nm^3/kg dry wood)	1.35	1.34
Composition (% vol./dry gas) CO	43.0	38.8
H_2	32.4	30.9
CO_2	23.9	27.4
CH_4	0.7	3.0
Thermal yield (%) LHV gas/LHV wood	69	69
Thermal losses (% LHV wood)	7	3
Tons potential methanol per ton dry wood *	0.49	0.44

* Comment : The difference in efficiency between the Le Creusot and the Clamecy pilot plant is due to :

1. The steam to wood ratio in case of Clamecy is five times higher than in case of Le Creusot. This is caused by pressure, in order to maintain the gas velocities in the fluidized bed.
2. The methane content in case of Clamecy (3 %), which is higher than in case of Le Creusot.

However, in case of Clamecy, these values are the result of some assumptions. They will have to be confirmed by the experiments which will allow to determine :

- the optimum operating pressure (with respect to the CO + H_2 yield), leading to the actual $\frac{\text{steam}}{\text{wood}}$ ratio.
- the optimum operating temperature under pressure.

6.3 Other test-runs

Before that 24 h official guarantee run, and after it, other test-runs were performed on the modified atmospheric pilot plant of a 400 kg/h wood capacity.

Particularly, in order to prepare the guarantee run, a 24 hours test was organized on February 1984, the 28 and 29th.

The results of that test run are summarized in the following table :

Wood feed-rate = 350 kg/h
Moisture = 21.5 %

Steam = 30 kg/h

O_2 fluid bed = 42 Nm^3/h
O_2 second reformer = 75 Nm^3/h

Uncontrolled air = 13.4 Nm^3/h
(with wood entry)

Argon tracer = 11 Nm^3/h

Fluid bed t = 720°C
Second reformer T = 1 380°C

O_2/wood = 0.61 kg/kg dry
Steam/wood = 0.11 kg/kg dry

Dry gas comp. (% vol.)

CO	= 37.0
H_2	= 30.9
CO_2	= 25.1
CH_4	= 1.6
Ar	= 2.9
N_2	= 2.6

Dry gas = 406.7 Nm^3/h
Steam = 186.0 Nm^3/h
Raw syngas = 592.7 Nm^3/h

Equivalent T of water gas shift equilibrium = 1 100°C

Thermal yield = 70 %

Carbon balance = 101.5 %
Hydrogen balance = 100.7 %
Oxygen balance = 100 %

Potential methanol =
0.49 kg/kg dry wood

After the guarantee run was performed, we made some other test runs to fill syngas bottles for testing, by LURGI, of methanol synthesis. For that purpose, 7 bottles of 10 m^3 syngas were filled, at a pressure of about 150 bar. These bottles and the compressor were then sent to LURGI Frankfurt. The bottles were filled when the CO + H_2 content of dry gas was roughly higher thant 65 %.

7 THE PRESSURIZED PILOT PLANT OF CLAMECY

The second phase in the development of the syngas from wood production process of CREUSOT-ENERGIE was started up in early 1984. The A.F.M.E. brought its financial support to the engineering and construction of the pressurized pilot plant of Clamecy (department of Nièvre). The pilot will be erected in that forestry region, on the site of a Rhône-Poulenc Chemical plant.

The aim of the project is to built and to experiment a pressurized gasification unit, of a 60 t/day dry wood capacity.

The main equipment of the pilot will be :

- the fluidized bed gasifier fed with wood chips, oxygen and steam,
- the secondary reformer fed with the raw gas coming from the gasifier and with secondary oxygen, to convert methane and higher hydrocarbons,
- the gas treatment (scrubbing-cooling...) allowing to burn the gas in a boiler.

The operating pressure should vary between 10 and 25 bar. The wood moisture will be about 15 %.

The schematic bloc-diagram of the unit is presented in figure 21.

The nominal operating conditions of the pilot are :

- pressure : 15 bar
- dry wood feed-rate : 2 500 kg/h
- moisture = 15 %
- O_2 flow-rate : 1 490 kg/h
- steam flow-rate : 1 000 kg/h

The expected characteristics of the syngas which is delivered at the battery limits are :

- raw gas flow-rate : 3 700 Nm^3/h
- dry gas flow-rate : 3 350 Nm^3/h
- dry gas composition (% volume) :

. CO : 39
. H_2 : 31
. CO_2 : 27
. CH_4 : 3

For this project, CREUSOT-ENERGIE and the A.F.M.E. are associated in a pool of Economic Interest named ASCAB ("Association pour le développement des carburants par la gazéification du bois", which means "Association for the development of substitute motor-fuels from wood gasification"). The pressurized pilot plant of Clamecy will be operational in 1986.

We hereafter present the main operating principles of that pressurized pilot plant of Clamecy.

7.1 Wood feeding and gasification

The preliminary dried wood is fed into the fluidized bed gasifier by means of a pressurized lockhopper system working as follows :

Wood chips are introduced in a lockhopper which is then pressurized with nitrogen up to the gasifier pressure. When pressures are identical, the bottom valve of the lockhopper is opened and wood-chips fall down into the meter bin. After being isolated again, the lockhopper is depressurized and then is ready for the introduction of a new wood charge.

Steam and oxygen injected at the bottom of the gasifier ensure fluidization of the inert material making the fluid bed.

The mixture of these streams lead to simultaneous thermochemical reactions in the refractory lined gasifier :

1 - Release of steam contained in biomass (moisture).
2 - Release of the volatile matters (pyrolysis), charged with CH_4, C_2H_4 and other hydrocarbons, which react with oxygen and steam to produce CO, CO_2, H_2, H_2O. Part of these volatile matters remain nonconverted (CH_4, C_2H_4).

3 - Combustion-gasification of the residual char which reacts with oxygen, steam and carbon dioxide to produce also CO and H2.

Finally, the gas leaving the fluidized bed gasifier is mainly composed of CO, H2, CO2, H2O and CH4.

Minor components of the raw gas are nitrogen, C2H4, C2H6, ashes and unconverted carbon.

The height of the fluid bed can be adjusted during the runs with a sand feeding system (lockhopper) and with an extraction device.

For the start up procedure, propane and air allow to preheat the gasifier and the secondary reformer.

7.2 Secondary reformer and gas cooling

The raw gas leaving the gasifier is introduced in the secondary reformer where it reacts at about 1 300°C with additional oxygen. So, residual methane and carbon are decomposed into CO and H2.

Ashes contained in the wood which are elutriated from the fluid bed melt and then are collected at the bottom of the secondary reformer, in the cooling water tank.

The gas is cooled down to about 200°C, by water injection in the cooling water tank. Then, the cooled gas enters the venturi scrubber.

Melted ashes coming from the cooling water tank are entrained in the water stream and then come into the flash tank.

7.3 Gas scrubbing and steam condensation

The raw syngas leaving the cooling water tank is charged with fine solid particles. It then passes through a venturi where fine droplets of water can catch solid particles, in order to transform the gas-solid in a liquid-gas separation. So, the gas charged with droplets comes into the scrubber.

The clean gas leaves the scrubber, saturated with water. Scrubbing water is recycled to the cooling water tank.

Then, the clean gas is cooled through a tubular heat exchanger where the main part of its water vapor is condensed, so that the calorific value of the syngas can be compatible with the downstream processing. During the first part of the experimental program, the downstream processing will consist of the combustion of syngas by the flare or in a boiler, or even in the woodchips dryer (partly). Condensed water leaving the gas cooler is collected in the separator and then recycled to the scrubber, in order to improve the gas-solids separation.

7.4 Process water streams

The excess water leaving the cooling water tank is injected into the flash tank, where it is expanded at about 3.5 bar to release dissolved and uncondensable gases. Flashed steam is then released at the stack.

After being cooled in an other heat exchanger, flashed water enters into the solids separator for settling.

From the bottom of the settler, the sludge (about 30 g/l) is pumped and conveyed to the process water treatment. The clear water leaving the solids separator is then recycled in a water tank where make-up water can be injected, depending on the needs of the process.

From the water tank, process water is injected into the cooling water tank and in the venturi for the gas treatment.

SCHEMATIC DIAGRAM - PRESSURIZED UNIT OF CLAMECY
CREUSOT-ENERGIE SYNGAS FROM WOOD GASIFICATION PROCESS
FIGURE 21

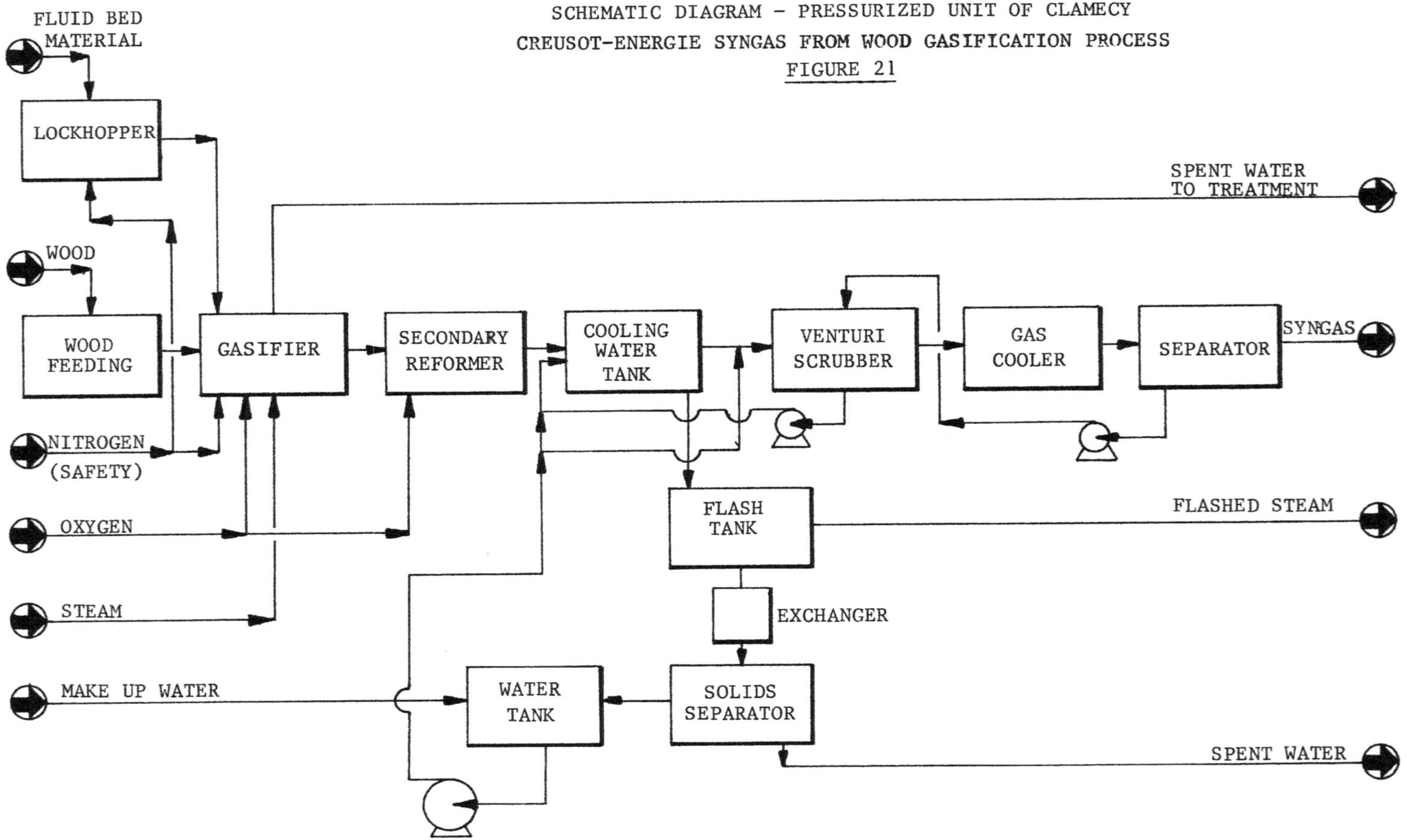

8 OUTLINE OF A COMMERCIAL PLANT

In that chapter are presented the different units which should enter in a methanol from wood plant. There are :

- Unit 010 : Wood storage and preparation
- Unit 020 : Wood drying
- Unit 030 : Gasification
- Unit 040 : CO shift conversion
- Unit 050 : Gas compression
- Unit 060 : Acid gas removal
- Unit 070 : Methanol synthesis and purification
- Unit 080 : H2 recovery (P.S.A.)
- Unit 090 : Spent-water treatment
- Unit 100 : Raw-water treatment - Cooling water - Fire protection water
- Unit 110 : Boiler feed-water pretreatment
- Unit 120 : Utilities (air for instruments, CO2, flares)
- Unit 130 : Methanol storage and handling
- Unit 140 : Oxygen production

8.1 Units 010 and 020 : Wood storage and preparation - Wood drying

Wood is supposed to come into the plant grinded, to the classical dimensions of paper-mill chips (less than 50 mm). Wood chips are stored, then dried in a battery of rotary driers. That technology is the most used, allows an homogeneous drying in spite of variations in chips dimensions. After drying, biomass contains 15 % moisture (on a raw basis).

For drying, hot air is generated in a wood kiln. We estimate that drying step, decreasing wood moisture down to 15 %, to consume about 14 % of total wood.

In fact, in a more energetically integrated methanol from wood plant, the part of wood for drying could be decreased.

8.2 Unit 030 : Gasification

That unit has been described in the previous chapter 7 "The pressurized pilot plant of Clamecy", § 7.1, 7.2 and 7.3.

8.3 Unit 040 : CO shift conversion

In that section, the gas mixture containing CO, CO_2, H_2 and H_2O is converted in catalytic reactors where the H2/CO ratio is adjusted at the required level. The chemical reaction involved is the water gas shift equilibrium : $CO + H2O \rightleftarrows CO2 + H2$.

Taking in account the H2 recovery (P.S.A.) on purge gas at the outlet of the methanol converter, the H2/CO ratio after the CO shift conversion is adjusted to 2.00.

Such a reaction can be performed in a single stage reactor, the condition is that temperature doesn't exceed 450°C.

8.4 Unit 050 : Gas compression

This is a classical problem. The purpose is to increase pressure at the methanol synthesis pressure requirement.

8.5 Unit 060 : Acid gas removal

In that section is eliminated most of carbon dioxide coming from the gasification and the CO shift sections.

A RECTISOL or SELEXOL process can be chosen, where CO_2 is absorbed by an appropriate solvent which is then recovered. RECTISOL seems more convenient in that application, since the solvent for absorption is methanol.

The process is a physical wash using an organic solvent (methanol) at temperatures below 0°C. In general, methanol is used for H_2S, COS and CO_2 removal.

8.6 Unit 070 : Methanol synthesis and distillation

A classical methanol synthesis process can be selected. The low-pressure ICI or Lurgi Process can be selected. The CO + H_2 mixture is converted into methanol between 50 and 70 bars.

Crude methanol is then purified by distillation, then stored.

Note : a new process, developped by I.F.P. (Institut Français du Pétrole) consists in the direct production, from a syngas composed of CO-CO_2-H_2, of an alcohol mixture extending from methanol to pentanol. The operation is performed in two steps : syngas, at 70 bars, pases first in a higher alcohol (> C2) synthesis reactor ; then, after having condensed water and alcohols, methanol is produced in a second catalytical area. The result of that double operation is a mixture of 85 % wt. methanol and 15 % wt. higher alcohols. The advantage of such a mixture is that it can be added in classical motor-fuels without any co-solvent. For that motor-fuel application, this process still in development appears very promising.

8.7 Unit 080 : H_2 recovery (P.S.A.)

The P.S.A. : "Pressure Swing Adsorption" process has been selected. Hydrogen contained in the raw purge gas from methanol converter is recovered at a roughly 99 % purity and then recycled before the compression stage.

8.8 Unit 090 : Spent-water treatment

The overall spent water of the plant is sent in a neutralization tank. It is then submitted to a three steps biological treatment before being sent in the draining system of the site.

Note : Environmental aspect of the commercial plant

Due to the high temperature at which the raw gas is processed in the secondary reformer, the syngas practically contains no organics : no tars, no oils, no phenols. In fact, the only really polluting component of syngas is cyanide CN^-, which is formed in the high temperature step. A specific treatment (with CARO-acid, or complexation under the form of ferrocyanides) will be added in the effluent treatment stage.

An important advantage of wood as raw material for syngas production is that it contains very low amounts of sulphur and chlorine. However, to protect the methanol synthesis catalyser against S and Cl, a zinc oxide guard is added before entering the methanol reactor ; traces of COS and H_2S are thus eliminated.

8.9 Units 100-140

Units 100 to 130 are classical ones.

In the unit 140, oxygen is produced by liquid air distillation. In some specific situations, oxygen should be produced by water electrolysis. In such a case, hydrogen should be added to the syngas, increasing the methanol yield per ton of wood ; in that case, the CO shift conversion unit should not be required.

9 CONCLUSION

As outlined above, much work has been performed during the four years of the development of the CREUSOT-ENERGIE syngas from wood process. Financial help of both the C.E.C. and the A.F.M.E. ("French Agency for Energy Savings") were important.

The results achieved with the atmospheric pilot plant of Le Creusot confirmed that the process is suitable for syngas production. The two stage gasification process leads, for a 15 % wood moisture, to a "CO + H2" dry gas content higher than 70 % on a volume basis. The thermal yield is about 70 %. The potential methanol is about 0.5 kg per kg dry wood.

In order to continue the development, the construction of the demonstration unit of Clamecy was decided in early 1984. The contribution of the A.F.M.E. to the project has been determinative. The demonstration unit of Clamecy will be based on the same two stage process, and will be operated under a 15 atmosphere pressure, at a wood feed-rate of 60 metric tons/day.

The purpose of that demonstration unit, the detailed engineering of which being currently in activity, will be to verify the ability of the process to be operated under pressure and also to experiment and put to the test the technological devices that would be applied on industrial units.

The beginning of experiments is planned for 1986. At the present time, the overall duration of the experimental program is eight months.

In a further step, that demonstration unit could be coupled to a methanol synthesis loop.

BIBLIOGRAPHY

- A survey of biomass gasification - Volume II - Principles of gasification Solar Energy Research Institute - July 1979

- Etude bibliographique sur la gazeification du bois - P. BELLEVILLE ; R. CAPPART - Université de Technologie de Compiègne - 1981

- Technology and Economics of wood residue gasification - Proceedings Tenth Texas Industrial Wood Seminar - Marche 13, 1979

- Steam gasification of wood in a multi-solid fluidized bed gasifier - H.F. FELDMANN - Battelle Colombus Laboratories

- Wood gasification in a fluidized bed - S. BECK - Department of Chemical Technology - Texas Tech. University

- Fluidized bed gasification of solid wastes and biomass : the C.I.L. Program - J.W. BLACK - Canadian Industries Limited

- Fast pyrolysis/gasification of lignocellulosic materials at short residence time - X. DEGLISE - Universite de Nancy

- La pyrolyse et la gazéification des déchets cellulosiques en couche fluidisée - J. SCHOETERS ; H. MASSON - Université libre de Bruxelles

- Fluidization engineering - KUNII - Levenspiel

- Handbook of Chemistry and Physics - 62th edition

- Chemical Engineer's Handbook - PERRY ; CHILTON - 5th edition

- Thermochemical processing of biomass - Edited by A.V. BRIDGWATER - Butterworths

- Energy from biomass - Serie E - Vol. 2 - Proceedings of the workshop on biomass pilot projects on methanol production and algae, held in Brussels - October 22, 1981

- Energy from biomass - Serie E - Vol. 5 - Proceedings of the workshop and EC Contractor's meeting held in Capri - June 7-8, 1983

- A.F. ROBERTS - Combustion and Flame, 14 (1970) - pp. 261-72

- K. AKITA - Report of Fire Research Institute of Japan (9) (1959) - pp. 1-44, 51-4, 77-83, 99-105

- A.J. STAMM - Ind. and Eng. Chemistry, 48 (1956) - pp. 413-17

- T. KUJIRAI and T. AKAHIRA - Scientific Papers of the Institute of Physical and Chemical Research (Tokyo), 2 (1924) - pp. 223-52

- S.L. MADORSKY, V.E. HART and S. STRAUS - Journal of Research NBS, 60 (1956) - pp. 343-54

- A.E. LIPSKA and W.J. PARKER - Journal of Applied Polymer Science, 10 (1966) - pp 1439-53

- W.K. TANG - US Forest Service Research Paper, FPL 71, 1967

- W.K. TANG and W.K. NEILL - Journal of Polymer Science, Part C, 6 (1964) - pp. 65-81

- C.H. BAMFORD, J. CRANK and D.H. MALAN - Proc. of Cambridge Phil. Soc., 42 (1946) - p. 166

- E.R. TINNEY, 10th Symposium on Combustion, Great Britain, 17-21 August, 1964

- A.M. KANURY and P.L. BLACKSHEAR - Combustion Science and Technology, 1 (1970) - pp. 339-55

- A.M. KANURY, PHD THESIS, University of Minnesota, March 1969

- H.S. KUNG and A.S. KALEKAR - Combustion and Flame, 20 (1973) - pp. 91-103

- A.F. ROBERTS and G. CLOUGH - Thermal decomposition of wood in an inert atmosphere, 9th Symposium on Combustion - Academic Press, New York, 1963 - p 188

- F.J. KANSA, H.E. PERLEE and R.F. CHAIKEN - Combustion and Flame, 29 (1978) - pp. 311-24

- J.VILLERMAUX and B. ANTOINE - Revue Génerale de Thermique (227) (1980), pp. 851-60

- J. VILLERMAUX, F. VERZARO and J. LEDE - Revue Générale de Thermique (227) (1980) - pp. 861-9

- F.L.BROWNE and W.K. TANG - Fire Research Abstracts and Reviews, 4 (76) (1962)

- F. SHAFIZADEH - Ind. and Eng. Chemistry, 20 (1981) - p. 482

- J.A. HAVENS - Thermal decomposition of wood, PHD THESIS, University of OKLAHOMA, 1969

- J. WIDELL - Acta Polytechnica, Chemical and Metallurgical Series, 1, (6) (1949) - p. 33

- P. BELLEVILLE, R. CAPART and M. GELUS - Pyrolysis of large wood samples - Applied Energy 16 (1984) - pp. 223-237

SYNTHETIC FUEL FROM WOOD VIA GASIFICATION IN THE CIRCULATING FLUID BED

Authors : Dr. P. Mehrling
: Dr. R Reimert

Contract number : ESE-P-067-D (b)

Duration : 21 months /
1st January 1982 - 30th September 1983

Total budget : DM 1,706,950.-- CEC Contr.: DM 853,465.--

Head of project : Dr. P. Mehrling

Contractor : LURGI GmbH / Department of Coal Energy

Address : LURGI GmbH
Gervinusstraße
6000 Frankfurt/Main

1. SUMMARY

Gasification of biomass yields a universally applicable intermediate or final product gas. From a plant operator's point of view, a gasification process has to be simple, effective and reliable, in addition to some other characteristics. To fulfil these requirements, Lurgi selected the Circulating Fluid Bed principle (CFB) for its new biomass gasification process. This process principle has been applied very successfully in calcination and combustion for more than a decade.

The applicability of CFB biomass gasification has been demonstrated in several pilot plant test runs. The test results have confirmed the expected process advantages, e.g. tar-free gas, high specific throughput, good partial-load behaviour, broad range of feed particle sizes. A report of the test trials and summary of the test results will be presented below.

A steam/oxygen mixture is used as the gasifying agent to produce synthesis gas. The synthesis gas consists mainly of CO and H_2, some CH_4 and CO_2 and a small amount of hydrocarbons. The suitability of the gas produced for methanol synthesis was verified by experiment.

Based on the pilot plant data on the one hand, and on extensive experience with pressure gasification, gas cleaning and gas processing plants on the other, plant concepts were established for methanol production. Cost estimates performed in parallel show under which conditions these processes are economically attractive.

TABLE OF CONTENTS

2. INTRODUCTION

Increasing fuel prices during the last decade have focused interest on the utilization of biomass as an additional source for the production of gaseous and liquid fuels. Furthermore, the predicted oil shortage has initiated governmental activities to subsidise the development of processes for the utilization of biomass.

At the present time, forestry and farm residues together with wood waste from the pulp and paper industries are the main potential sources. In addition, fast-growing biomass from energy plantations is expected to play its role in the near future.

Considering the heterogeneous character of these feedstocks, gasification appears to be the right process to produce a feasible liquid or gaseous fuel of general specification.

The gasification process to be developed under these conditions should meet the following requirements which are not yet fulfilled by other processes available:

- handling a broad spectrum of biomass
- allowing for a high throughput in the range of 100 000 m^3_n/h
- producing a synthesis gas suitable for direct methanol synthesis without further reforming
- environmentally acceptable

A gasification process working according to the Circulating Fluid Bed principle (CFB) promised to fulfil the requirements mentioned above and was therefore selected for development. Lurgi's decision was based on long experience with this process principle since its introduction in the alumina industry fifteen years ago.

2.1 Aim of the Project

In particular, the aim of the development project reported on in the following was:

- Demonstration of wood gasification in a CFB-gasifier.
- Determination of the influence of independent parameters on gasification performance.
- Compilation of a process scheme "Methanol from Wood", including estimate of total investment cost and
- Estimate of production cost for methanol

2.2 CFB-Gasification - Process Principle

The CFB process may be placed in the transition range between the stationary bubbling fluidized bed with defined surface and the forced pneumatic transport reactor (Figure 1).

A gas flowing through fine-grained solid material with increasing velocity will initially generate a fluidized bed of the classical form with defined surface. It can be compared with a boiling liquid (condition A). At this stage, only relatively few solid particles are entrained with the gas; they are separated in a cyclone and either discharged from the reaction system or recycled to the fluidized bed.

As the gas velocity increases, the fluidized bed expands until the solids are almost uniformly distributed over the reactor height. The quantity of solids entrained by the gas increases. The solids are separated in the recycle cyclone and recycled to the reactor, creating the circulating fluidized bed (B). This makes the recycle cyclone a permanent constituent of the reactor system.

The external circulation is accompanied by an inner recirculation of material in the reactor due to constantly changing densely packed strands and clusters. This leads to an intensive mixing of gas and solids. In the diagram, this fact is characterised by the increase in relative slip velocity between gas and solids. Together with the solids recycle, the inner circulation ensures a uniform temperature across the overall circulating fluidized bed reactor.
The CFB operates with optimum effect in the range of maximum possible slip velocity for the given particle spectrum. This causes extremely good heat and mass transfer which enables a rapid heating of the feed material, combined with an excellent reaction rate at constant temperature in the circulating fluidized bed system as a whole.
Lurgi's experience with CFB goes back to the sixties, when the development of this process was started. In the seventies, this process was introduced in the alumina industry for calcination. Up to now, more than 35 industrial plants have been installed for different endothermic heterogeneous gas-solids reactions working according to the CFB-calcination technology.

The success of these plants led to the Circulating Fluid Bed being used for combustion of coal in industrial and utility steam generating facilities as well. Lurgi conducted extensive tests in its two pilot plants in Frankfurt and constructed its first commercial-scale CFB coal combustion plant in mid-1982. This plant with a capacity of 84 mw_{th} was built for VAW (Vereinigte Aluminium-Werke) in Luenen, West-Germany. Three other plants are currently under construction.

Since the technology was applied with great success to calcination - an endothermic process - and to combustion - an exothermic process -, application to gasification, which is an exothermic process seemed a logical extension. Nevertheless, pilot scale tests were required to confirm the expectations.

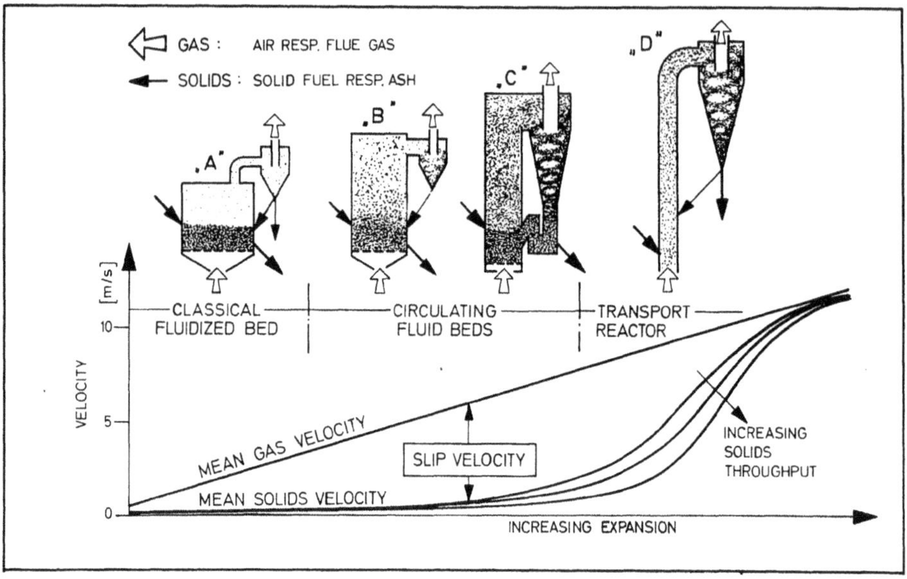

FIGURE 1 Basic Systems for gas/solid reactions for fine grained solids

	BEECH WOOD	PINE WOOD	POPLAR WOOD	RECYCLE WOOD
ASH (dry) (%wt)	3.4	3.0	14.5	5.3
ELEMENTARY ANALYSIS (daf)				
C (% wt)	48.0	49.0	51.5	50.8
H (% wt)	5.6	5.6	6.5	5.4
O (% wt)	45.9	45.2	41.6	43.4
N (% wt)	0.4	—	0.3	0.3
S (% wt)	0.1	0.2	0.1	0.1

TABLE I Elementary analysis of wood used during gasification tests

3. PRETEST PHASE

The pretest phase mainly comprised three activities:

- characterisation of feed material,
- design of pilot plant modification, engineering and construction. (The pilot plant had been designed and constructed to develop calcination and therefore had to be modified for gasification purposes) and
- definition and compilation of a detailed test programme.

3.1 Characterisation of Feedstock

With respect to its calorific value, biomass is a low grade fuel, characterized by high oxygen and hydrogen contents and a relatively low carbon content. A typical composition in molar ratios is:

$$C_6\ H_{8,82}\ O_{4,14}\ N_{0,03}\ S_{0,005}$$

These renewable resources, containing lignite and cellulose, are closely related to fossil fuels like coal, lignite and peat. In a graph showing the molar ratios of H/C versus O/C (Fig. 2), the hatched region encloses the fossil fuels derived from lignite and cellulose. The isocalorific lines plotted in addition indicate the decreasing heating value with increasing O/C.

The feedstocks used in gasification tests have been analysed in LURGI's laboratory. The composition as determined is given in Tab. I.

3.1.1 Determination of the Reactivity of Feedstock under Ambient and Elevated Pressure

Although feedstock stoichiometry is an important figure, it is not the only factor determining gasification behaviour. Reactivity of feedstock is another. The determination of reactivity can be performed at ambient or elevated pressure. The Lurgi method for the characterisation of feedstocks covers both. The procedure can be described as follows.

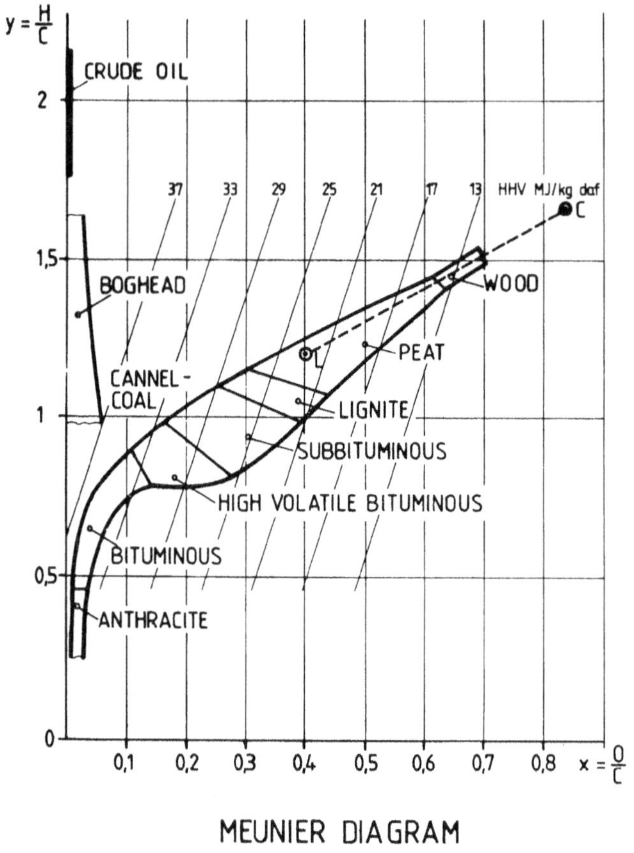

FIGURE 2 Molar ratio H/C versus O/L and HHV of genuine solid fuels "Meunier Diagram" (2)

	Reactivity (ml CO/g.s)	rate constant $cm^3/g.s$
Pressure	ambient	ambient
Temperature	800 °C	800 °C
Beechwood	0,370	1,15
Pine	0,602	2,33
Poplar	0,535	1,91
Spruce	0,514	2,0

TABLE II -1 Reactivity and rate constant of wood char versus CO_2 at ambient pressure

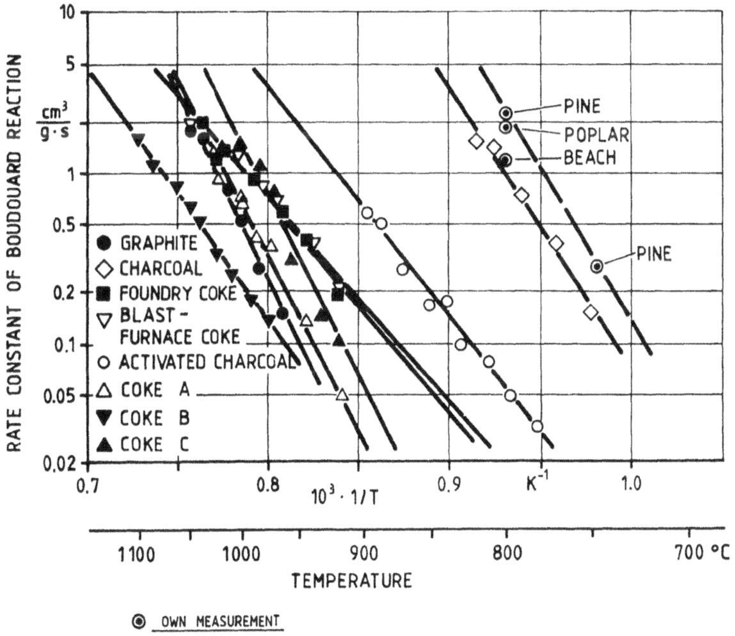

FIGURE 3 Rate Constants k_m of different fuels gasified at different temperatures (3)

The feedstock to be investigated is carbonized under the condition selected for the following reactivity tests. The char is then transferred into a tubular reactor designed for tests under ambient or elevated pressure, heated and partially converted in a CO_2 gas stream for approximately 1 h. The relative conversion rate, the reactivity (ml CO/g.s), or the rate constant, is calculated by integrating the CO produced during the reaction time.

These tests were conducted both under atmospheric conditions and at a pressure of 20 bar. It must be pointed out that the determinations in question are methods of determining relative reactivity values which are technically applicable, and not scientific kinetic measurements.

The results of the reactivity measurements under atmospheric conditions are listed in Table II-1.

The reactivity of lignite, which is of the same magnitude as the reactivities of the woods investigated, is indicated for comparison. In addition, a classification of the experimentally-determined reactivities in relation to those of other solid fuels is possible with the aid of an Arrhenius plot (Fig. 3). The high reactivity of the wood compared with other solid fuels is clearly evident.

The results of the measurements at elevated pressure are listed in Table II-2 together with the value for lignite. Here too, the reactivities of the woods investigated are of the same order as the reactivity of lignite. Both the measurements at ambient pressure and those at elevated pressure only provide relative values due to the described characteristics of the measuring methods. The measurements at elevated pressure, in particular, are not directly comparable with those at ambient pressure. However, the fact that the values determined for wood are of the same order as the corresponding values for lignite both at ambient pressure and at elevated pressure shows that the well-known influence of pressure on the CO_2 reactivity of coal, i.e. the reactivity increases with the pressure, also applies to wood.

rel.	Reactivity (ml CO/g.)
Pressure	20 bar
Temperature	800 °C
Beechwood	0,162
Pine	0,182
Poplar	0,204
Lignite	0,138

TABLE II-2 Reactivity and rate constant of wood char versus CO_2 at elevated pressure

	Engineering	Fabrication	Installation
P & I Diagram	x		
Gasification agent Supply and distribution	x	x	x
Wood feeding system	x	x	x
Gas scrubbing system	x	x	x
Safety-system	x		x

TABLE III CFB wood gasification
- Pilot Plant Modifications -

3.2 Description of Pilot Plant and Plant Modifications

The CFB pilot plant used in the test, originally designed for calcining aluminum hydroxide, had to be adapted to the specific requirements of the gasification process. A complete new process flow-diagram was developed, from which a schedule of equipment to be modified or newly fabricated and installed was prepared. A summary of these activities is given in Table III. Finally, after installation and commissioning of new and modified equipment, the pilot plant was ready for gasification tests.

A description of the pilot plant and of plant operation is given below; a process flow diagram can be seen in Fig. 4.

Prepared wood enters the wood bin B-101 and then wood lock B-102. From the lock, the wood passes by gravity into storage bin B-103. It is conveyed and proportioned by the cooled screw feeder H-101.

The gasification agent, either air or an oxygen/steam mixture, is preheated if necessary and supplied to reactor D-201. The gasification reaction between wood, coke and gasification agent takes place in gasifier D-201. Particles entrained by the raw gas are separated in the hot cyclone F-401 and recycled directly into the reactor. The ash is withdrawn by screw H-301 and discharged into ash bin B-301.

The raw gas produced by gasification is dedusted in hot cyclone F-403 and scrubbed and saturated in downstream scrubber W-401 with water recycled via pump P-401.

The main product gas flow is burned in combustion chamber D-401 and discharged to the air via offgas booster V-401. A side stream of the clean cooled product gas can be branched off and bottled for further use, e.g. demonstration of methanol synthesis.

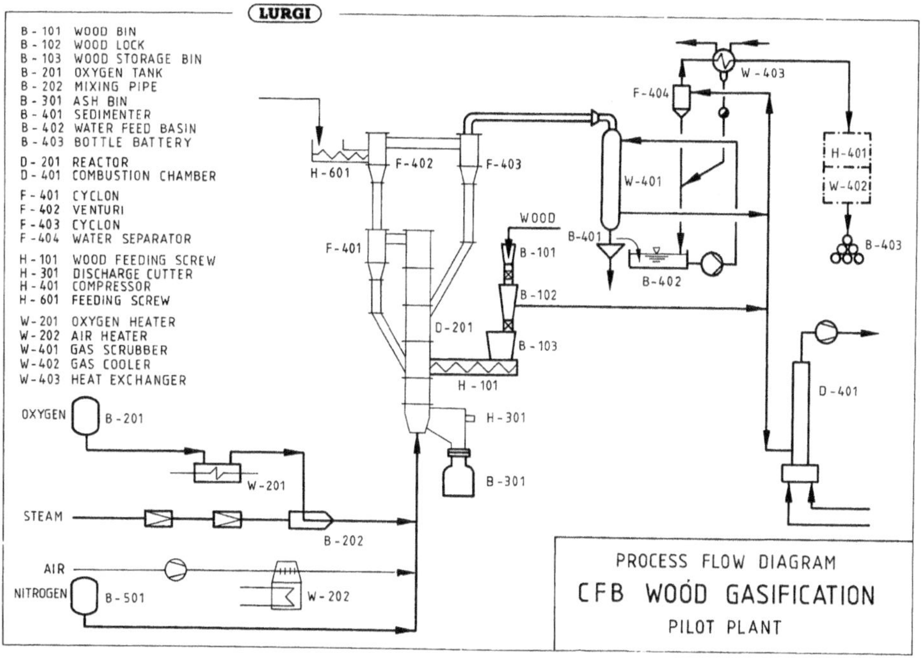

FIGURE 4 Process Flow Diagram of CFB-Wood-Gasification

KIND OF WOOD	BEECH-, PINE-, POPLAR-, SPRUCE
GRAIN SIZE	0 - 30 (70) MM
GASIFICATION AGENT	OXYGEN / STEAM
TEMPERATURE	630 - 800°C
PRESSURE	ATMOSPHERIC
BED MATERIAL	WOOD - CHAR
THROUGHPUT	UP TO 250 KG_{DAF}/H

TABLE IV Test parameters varied during CFB-Pilot-Tests

4. GASIFICATION TESTS

The pilot tests, which were performed for the development of CFB biomass gasification, were carried out in a series of 3 test campaigns. Table IV summarizes the parameters varied during these tests. The total operation time was approximately 1500 h, including a guarantee run of about 50 h. During these tests the feasibility of CFB biomass gasification was proven and a set of design data established.

4.1 Description of Test Runs

The test runs were subdivided into three series. The first was started is January 1983. The gasification was carried out at a relatively high gasification temperature of 1050 °C. It was found that due to this high temperature, sticking and agglomeration occured, even with Al_2O_3 as an inert bed material.

To avoid these problems, during the second test anthracite was used as an "inert" bed material. Then it was possible to reach gasification temperatures of 920 °C without operational problems and a continuous test run of 5 days was performed. However, during these tests a constant anthracite flow had to be added, which obviously indicated that anthracite did not act as an inert bed material but was gasified together with the wood even at a relatively low reaction temperature.

As this simultaneous gasification of anthracite was undesirable, particularly with a view to the subsequent evaluation of the tests, and the addition of an inert material is in any case not worthwhile in the operation of a future commercial plant for logistical reasons, an attempt was made to form the Circulating Fluidized Bed solely from wood char produced in the reactor itself. On the basis of previous experience, an appropriately low temperature had to be selected.

During the following third test run, optimum results were achieved. Due to the high reactivity of the wood char, the maximum reaction temperature was nearly 800 °C, the optimum temperature 750 °C. The guarantee run was performed under these conditions.

To summarize the operational experience reported above, it can be stated that after the first tentative tests, two run periods of 100 and 50 h of continuous operation were performed, showing the feasibility of the process.

4.2 Assessment of test results

The basis of the evaluation of the test results was the data, both independent and dependent parameters, collected during test runs. The independent parameters of CFB gasification, the mass flows of wood, inert bed material, steam and oxygen, have to be adjusted to reach the targeted reaction temperature.

The mass flow of wood, inert bed material and bed discharge were measured by weighing. The flow rate of steam and oxygen were measured by orifice, while the flow rate of product gas was calculated taking the carbon balance as a basis. The product gas was analysed continuously by mass spectroscopy and additionally by gas chromatography. Furthermore, the moisture and dust contents of the product gas were determined by isokinetic measurements.

A summary of the analytical methods applied during the tests is given in Table V.

The data collected as described above were used for the calculation of mass and heat balances. The most important results of these assessment are described in the following sections.

Carboniferous Solids Analysis	Method
Proximate Analysis	
Moisture	DIN 51718 II
Ash	DIN 51719
Ultimate Analysis	
C	ASTM D 3178 II
H	(Combustion)
N	DIN 51722 (Dumas)
S Combust.	ASTM D 3177
Cl	ASTM D 3177
O Diff.	by Difference
Gas Liquor	
pH	Apha 424, pH-meter /15/
H_2S/Sulfide	Apha 428D Ccl acetate method or potentiometric titration
NH_3 free, total	Apha 418 D, destillation + acidimetric filtration
Phenols	LURGI "Koppechar", equiv. to Apha 510
Gas-Analysis	
Continuous Analysis of Product-Gas	
Analysis	Method
H_2, CO, CH_4, C_2H_6	Process-Mass-Spectrometer
CO_2, N_2, O_2, H_2S	Perkin-Elmer MSA 1200
Gaschromatographic Analysis of Product Gas	
Standard GC	Hewlett-Packard HP 5880 A
N_2, O_2+Ar, CO, CO_2, H_2	3-column GC, 24 ft x 1/8"
HC_4, C_2H_6, C_2H_4	20% Sabaco nitrate
C_3, C_4	ast Chromosorb P AW 80/100;
	10 ft x 1/8" Molsieve 13 x 60/80 and 6 ft x 1/8" Porapax a 80/100 Detection HCD

TABLE V

Analytical Methods

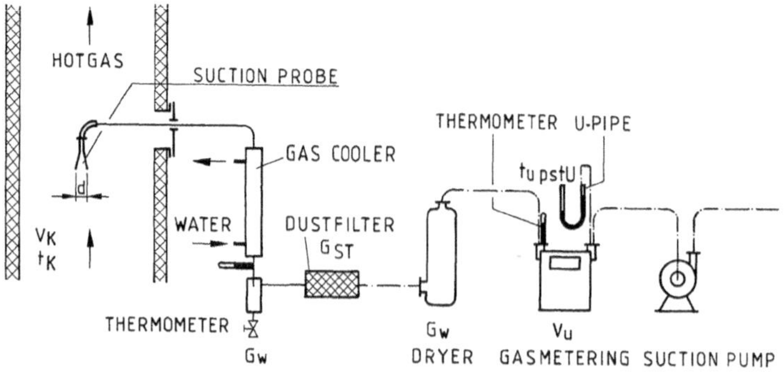

FIGURE 5 Isokinetic measurement of fines content and gas liquor in product gas

4.3 Quantitative Results

The quantitative and qualitative results are discussed in the following sections. Table VI gives quantitatively selected production and consumption figures, product gas composition and efficiencies of the test runs. A detailed analysis of the influence of independent parameters on CFB gasification is given in section 4.4.

The figures shown in table VI indicate generally high efficiencies and high carbon conversion. Besides beech, spruce, poplar and pine wood were gasified. As these types of wood were not available in dry condition, they had moisture contents between 40 and 65 % wt.a.r., leading to a product gas with a CO_2 content of up to 50 %.

Considering the gas quality, it can be seen that optimum results were achieved at a temperature range between 720 and 750 °C with wood char as a bed material. In this temperature range, high CO and H_2 contents and high efficiencies were achieved. The product gas also comprised moderate amounts of CH_4 and C_2-C_3 hydrocarbons, allowing the gas to be routed into methanol synthesis without reforming. In addition, the analysis of gas liquor, collected by cooling a product gas side stream, indicates that the gas is tarfree, whereas for stationary fluidized beds, tar contents between 400 and 6000 mg/m³n have been reported (4). Table VII shows the gas liquor analysis of the product gas of the gasification tests.

TABLE VII Composition of gas liquor (4)

Reaction Temperature	750°C	
pH	9	
Phenols	1,600	mg/kg H_2O
Tars	tar-free	"
Fatty acids	100	"
TOC	1,800	"
COD	6,000	
H_2O	0,2	kg/m³n

FEED (TYPE OF WOOD)	BEECH				
GRAIN SIZE (MM)	5 - 15	5 - 15	10 - 30	1 - 10	1 - 30
MOISTURE	21,3	17,0	21,0	22,5	21
TEMPERATURE	714	750	774	870 2)	923 3)
H_2 Vol %	33,4	32,3	30,8	20,0	29,8
CO "	30,0	32,4	32,9	24,8	37,1
CO_2 "	28,5	25,7	25,4	42,5	24,6
CH_4 "	4,3	4,7	5,2	8,8	6,1
C_2H_4 "	0,7	0,7	N.D. 4)	N.D.	N.D.
C_3H_6 "	0,6	0,6	N.D.	N.D.	N.D.
N_2	2,3	3,3	5,0	3,9	2,5
x H_2O KG/M_NDRY	0,3	0,2	0,3	0,4	0,3
H_2S PPM	----------10 - 20-------------------------				
RAW GAS M_N/KG DAF	1,41	1,40	1,32	1,13	1,44
OXYGEN CONS. M_N/KG DAF	0,32	0,32	0,32	0,46	0,30
STEAM 1)/OXYGEN KG/M_N	0,96	0,75	0,95	0,72	0,96 5)
COLD GASEFFICIENCY %	83,1	86,5	72	56	79
CARBON CONVERSION %	99	99	98	98	98
q_{spec} KJ/KG DAF	561	523	550	1280	385

1) STEAM = STEAM ADDED + MOISTURE

2) AL_2O_3 BED MATERIAL ADDED

3) 16 KG/H ANTHRACITE ADDED REDUCED THROUGHPUT

4) NOT DETERMINED

5) HIGH STEAM VALUE

6) q_{spec} SPECIFIC HEAT LOSSES

TABLE VI Selected results of gasification tests - specific Production - and Consumption Figures -

4.4 Influence of Independent Parameters on Gasification Performance

As previously mentioned, the influence of the most important parameters determining the gasification behaviour was also studied.

Temperature, moisture or steam addition and specific heat losses of the plant were found to be of major influence. The type and size of the wood are rather less important. In the discussion later on, wood moisture and steam in the gasification agent are regarded as parameters of the same kind and summarized as steam/oxygen ratio. Heat losses are, of course, something very specific to the individual plant. However, since heat losses have a strong influence on gasification behaviour, they had to be carefully examined for the purpose of data extrapolation.

The range of reasonable gasification temperatures is limited

- at the lower end by the production of undesired hydrocarbons, and
- at the upper end by the reactivity of the char produced within the gasifier.

For the Circulating Fluid Bed process, this range was found to be between 700 and 800 °C. Only a few C_2^+ hydrocarbons and traces of fatty acids and phenols were found in this temperature range and, in particular, no tars were produced (see also Table VII).

When applying higher temperatures, e.g. 1000 °C, the flow regime of fluidized beds cannot be maintained any longer, due to the higher rate of reaction. The char will then behave as in an entrained bed and will not circulate. Hence, an inert bed material would be necessary, whereas the CFB operates in principle without any additional bed material.

Figure 6 shows the dependence of the molar ratio of H_2/CO in the raw gas on the steam/oxygen ratio in the gasification agent, indicating an increase in H_2/CO content with increasing steam/oxygen ratio content.

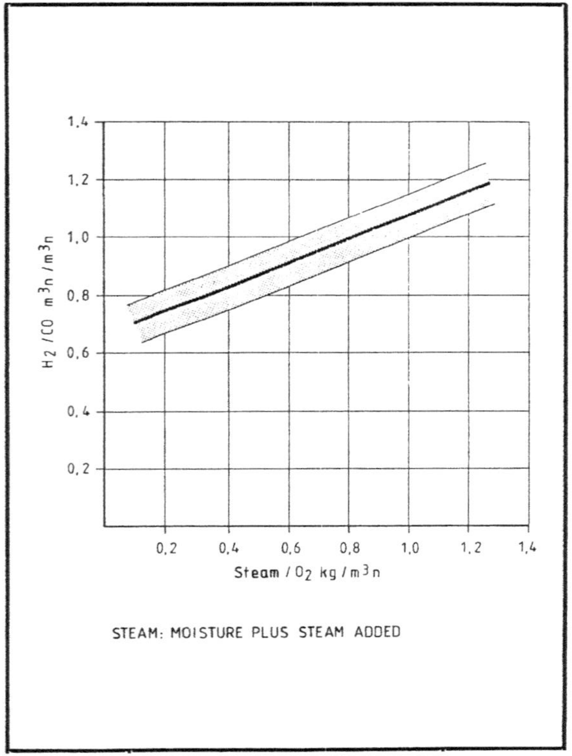

FIGURE 6 Dependence of H_2/CO ratio of CFB product gas on Steam/oxygen ratio of gasification agent

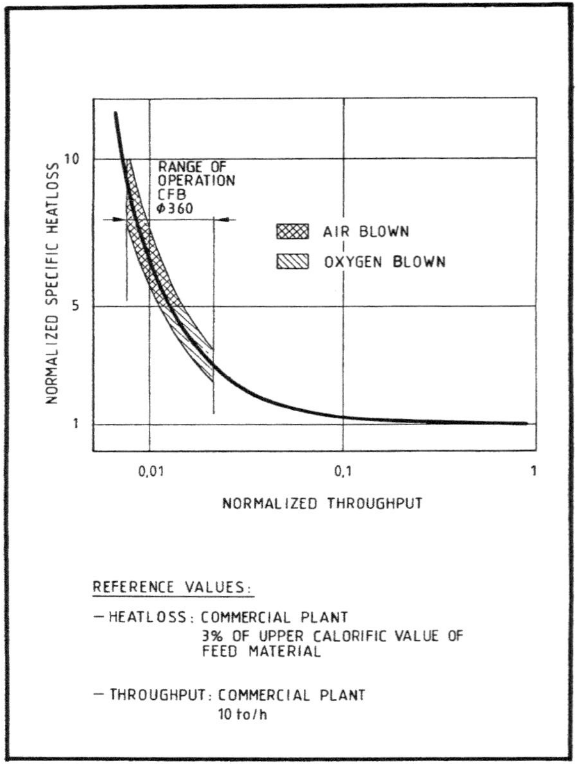

FIGURE 7 Effect of normalized throughput on normalized specific heat loss

These effects of temperature and wood moisture are typical for the process itself and independent of the plant size. In contract to these parameters, heat loss depends on the individual plant. For a more general formulation, the specific heat losses and throughputs of the pilot plant were normalized with data valid for a commercial-size plant (Table VIII). As a result, Figure 7 shows a graph of normalised specific heat loss $q^N_{spec.}$ as a function of normalised feed rate m^N_{Fuel}, indicating a strong dependency with low feed rates and a decreasing dependency with increasing feed rate.

The influence of normalized specific heat loss on plant performance can be seen in Figure 8. The decreasing amount of $CO + H_2$ is parallel by an increasing CO_2 content in the product gas.

These dependencies have to be taken into consideration when extrapolating pilot plant data towards those of a commercial-size gasifier. Results of these calculations are given in section 7.

Qualitatively, the test results can be summarized as follows:

- Wood gasification in a CFB reactor is feasible.
- High carbon conversion results in favourable specific consumption and production figures.
- High specific throughput allows for high-capacity plants.
- The small amount of methane makes the gas suitable for direct routing into synthesis processes without previous or simultaneous reforming.
- Tar-free gas indicates a high environmental acceptability of the CFB gasification process.
- No inert bed material needs to be added.

SPECIFIC HEAT LOSS

$$q_{\text{HEATLOSS SPEC.}} = \frac{q_{\text{LOSS PLANT}}}{\dot{m}_{\text{FUEL}}\ \text{HHV}_{\text{FUEL}}}$$

NORMALIZED SPECIFIC HEAT LOSS

$$\frac{q_{\text{HEATLOSS SPECIFIC}}}{q_{\text{HEATLOSS SPECIFIC OF COMMERCIAL PLANT}}} = q^{N}_{\text{SPEC}}$$

NORMALIZED THROUGHPUT

$$\dot{m}^{N}_{\text{FUEL}} = \frac{\dot{m}_{\text{FUEL}}}{\dot{m}_{\text{FUEL COMMERCIAL PLANT}}}$$

REFERENCE VALUES

$q_{\text{LOSS SPEC. COMMERCIAL PLANT}}$ = 3% OF HIGHER CALORIFIC VALUE ~700 kJ/kg DAF

$\dot{m}_{\text{FUEL COMM.}}$ = 10 000 kg/h

TABLE VIII Normalized specific heat loss and normalized throughput

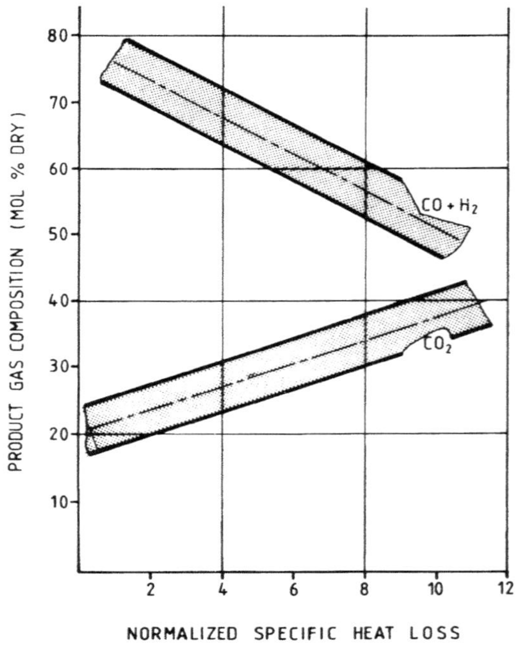

FIGURE 8
Influence of normalized specific heat loss on product gas quality of CFB wood gasification

5. COMPARISON WITH STATIONARY FLUIDIZED BED

One of the objectives of the project was to demonstrate plant operation of the CFB-process under the regime of a stationary fluidized bed.

Due to overall plant capacity and normalized specific heat losses of the test plant, it was not possible to run the plant at a sufficiently low throughput and only one short test run could be performed. Consequently, the comparison needs to be based mainly on data taken from literature.

For the purpose of comparison, the general characteristics of the Circulating Fluid Bed and stationary fluidized bed are summarized in Table IX.

	stationary fluidized bed	CFB
Distribution of fines in the reactor	high bed density at bottom	homogeneous
Temperature profile, radial	hot spots	homogeneous
axial	drop at bed surface	homogeneous
Feeding point	above bed surface	bottom of reactor
Reactor height	moderate	moderate
throughput	moderate	high

TABLE IX Qualitative comparison between CFB and stationary fluidized bed

The gas composition expected in the latter case should be close to the compositions achieved in stationary fluid bed gasification. In table X, data for a stationary fluidized bed are summarized together with data of the LURGI CFB gasifier at operation at normal and low gas velocity.

	T °C	CH_4 Vol %	C_2H_4 Vol %	higher hydrocarbons	Literature cited
HTW		approx. 6,5 1)			6
Mino		17 2)			7
Creusot Loire	800	10 - 12	2.5 - 3	0.1 - 0.5 3)	8
Batelle	750	13.6	1.7	1.0 4)	10
CFB	750	4.7	0.7	0.9 4)	
CFB 6) at low gas velocity	700	9.0	n.d.	0.1 5)	

Remarks
1) with secondary oxygen
2) at 30 bar
3) C_2H_6 (C_3H_n not determined)
4) C_2H_6 + C_3H_n
5) C_2H_6 (C_3H_n not determined)
6) Throughput 50 % of standard value

TABLE X Comparison of gas composition of stationary fluidized bed, CFB and CFB in operation at low gas velocity

The characteristic differences in bed density, solid partile distribution, temperature profile and feeding point of stationary fluidized bed result in an increase of CH_4, higher hydrocarbons and tars (Table X).
A reduction of throughput in a CFB gasifier is parallel by decreasing gas velocity in the reactor and consecutive changes in solid particle distribution. Therefore, it should be expected that the reaction conditions will come close to the conditions of a stationary fluidized bed and a change in gas composition should also be expected. Figures achieved in CFB gasification (Table X) at reduced throughput confirm these expectations.
Besides, this short tests with reduced throughput demonstrated the capability of the CFB reactor to be operated at stationary fluidized bed conditions. However, some features of the CFB are less apparent.

6. Methanol from Wood via CFB Biomass Gasification

With the pilot tests presented in section 4, a sound basis was attained for commercialisation of CFB biomass gasification. One of the potential applications of this process is the production of liquid fuels via methanol synthesis. For a better understanding, Figure 9 shows a simplified block flow diagram of this process.
For the time being, two types of methanol fuels are available as fuel additives, chemical grade methanol (Fig. 9) and fuel methanol. Whereas chemical grade methanol is produced in standard processes such as Lurgi Low Pressure Methanol Synthesis, a modified Lurgi process has to be applied (12) to the production of fuel methanol. Table XI summarises the most important characteristics of both kinds, indicating that especially in the case of fuel methanol produced from a syngas derived from CFB-biomass gasification, the plant configuration becomes less complex and an increase in thermal efficiency may be expected. As the fuel methanol process is completely new, it was not the basis of the design of the methanol-from-wood plant presented here, but shows a potential for further applications.

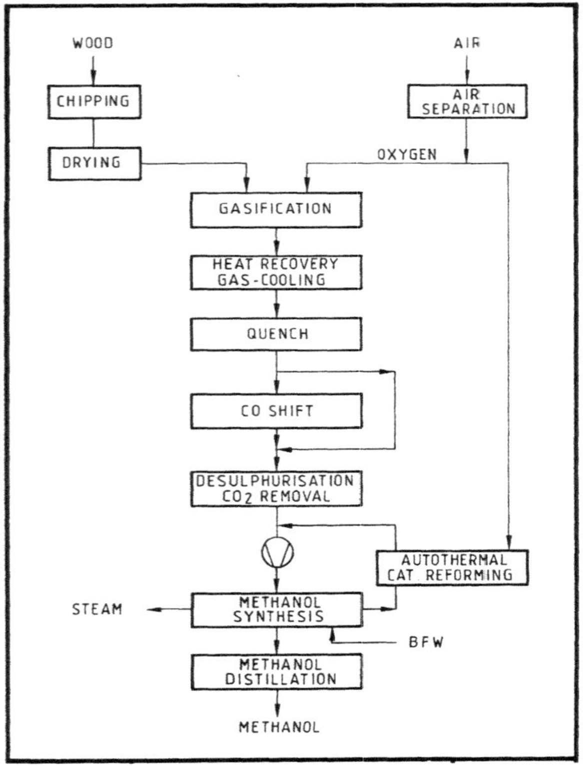

FIGURE 8 Block Flow Diagram: Methanol from Wood Plant

	Chemical Grade Methanol	Fuel Methanol
Product gas CFB		
H_2/CO	~ 1	~1
Syngas composition		
H_2/CO	2,4 - 2,5	1
Product composition		
CH_3OH	99	53
higher Alcohols		
C_2-C_5		up to 21
C_6+		up to 21
Application	chemicals, fuel fuel additive	octane booster (Fuel additive) (Fuel)
Advantages of syngas from CFB		no shift conversion less complex higher efficiency

TABLE XI Characteristics of chemical grade Methanol and "Fuelmethanol"

The methanol-from-wood plant presented below is based on the standard Lurgi low pressure methanol process. In detail, the following topics are dealt with:

- o demonstration of methanol synthesis from syngas derived from the CFB biomass gasifier
- o mass and energy balances of "methanol-from-wood plant"
- o detailed process description
- o cost considerations.

6.1 On-line Demonstration of Methanol Synthesis using Syngas* derived from Wood

To demonstrate the possibility of methanol synthesis with syngas derived from wood, a side stream of CFB product gas was branched off, compressed and bottled during the long-duration run (chapt.: 3.2.4). This gas sample of approx. 60 m^3n was conditioned in a shift conversion unit, purified in a Rectisol plant and routed to the laboratory synthesis loop - a tubular reactor with dimensions equivalent to one tube of the LURGI-low pressure methanol synthesis unit (7).
The synthesis reaction took place at standard operating conditions and at a pressure of 50 bar.
The composition of the feed gas for the methanol loop and the test results are given in Table XII.
With regard to data extrapolation, it has to be borne in mind that the test results reported here were achieved under operating conditions that had not been optimized. But due to the process know-how it is possible to calculate the expected figures from a commercial plant serving as a basis for the engineering of the "methanol-from-wood plant."

---

*) Demonstration of methanol synthesis was done as a service for EC dir. 12 under contract ECE-P-071-D(B)

Operating Conditions Standard

Gas Composition		Feed gas	purge gas
CO_2	% by vol.	1.48	3.0
CO	"	24.71	7.2
H_2	"	59.33	42.6
CH_4	"	6.11	20.5
$N_2 + C_2+$	"	8.37	26.4
CH_3OH	"	0	0.3

Test Results

Conversion of CO_2	40 %
" " CO	91 %

Specific Production Figures

Methanol:	0,33	kg/m³n feed gas
	0,35	kg/kg DAF wood

TABLE XII Demonstration of methanol synthesis using wood-derived syngas - gas composition and production figures -

6.2 Commercial-size "Methanol-from-Wood Plant" based on Lurgi CFB Gasification Process

Engineering and production data for the plant were derived and cost factors considered in order to investigate the question of whether the costs of methanol from wood can compete with methanol from fossil sources.
Under this aspect, the principal design data were selected as follows:

Plant Size	:	1,500 t/d
Feed	:	Beechwood
Steam/Energy	:	self-supporting, import of el. energy
Location	:	industrialized country (Europe)
Battery Limits	:	grass roots
Environmental impact	:	conforming to German standard for coal conversion plants

TABLE XIII Capacity, battery limits of "methanol-from-wood" plant.

Taking these data as the basis, engineering data for the complete methanol-from-wood process were then estimated. As shown by the simplified BFD (Fig. 8), the main components of this process are:

- o CFB gasification, including wood grinding and drying
- o Air separation with steam drive
- o Heat recovery and cooling (dedusting)
- o Shift conversion
- o Gas purification
- o Methanol synthesis including
  - Methanol distillation
  - Autothermal catalytic reforming

Whereas CFB-gasification has been newly developed as described above, all the downstream processes are proven and commercially available technologies. Therefore, only extrapolated production data of the pressurized version of CFB gasification will be discussed in detail below.

6.2.1 CFB Gasification - Pressurized Version
- Consumption and Production Figures -

The advantages of pressurized gasification units incorporated in large methanol plants have been reported elsewhere. In the case of CFB biomass gasification, a reaction pressure of 30 bar is taken as a design basis.

To compile performance data for such a CFB gasifier, the influence of pressure on gasification has to be estimated. This influence has not been yet very well analysed and the small-scale on regarding reactivity could not close this gap. But an analysis of data taken from the literature suggests that the amount of methane and higher hydrocarbons is mainly influenced by the thermal stability of primary pyrolysis products.

As the residence time of the pressurized CFB plant is about 5-7 times higher than during pilot tests in the atmospheric CFB plant, the methane content is considered to be about 10 % less than was determined during the pilot tests.

Table XIV gives the complete results of the extrapolation. These production and consumption figures were taken as a design basis in order to compile mass and heat balances for the methanol-from-wood process. A more detailed process flow diagram of the gasification unit is shown in Fig. 10.

6.2.2 Process Description

- Gasification section (Fig. 10)

  The main units of this section are

  - wood crushing and drying
  - CFB gasifier
  - Waste heat recovery, dedusting and cooling

  Wood is taken from a storage pile and routed via crushing and drying units into the wood feeding system, from where it is conveyed and proportioned into the gasifier via a lock hopper system and a screw feeder. The gasification agent is taken from an air separation unit. The product gas is cooled down, dedusted and saturated for further processing in the shift conversion.

- Downstream processing (Fig. 11)

  The cooled and dedusted raw gas is conditioned and, in particular, the H_2/CO ratio adjusted to the requirements of the methanol syn-loop (16) in a two stage clean gas shift conversion. The shifted gas is then routed to a purification unit (a standard LURGI Rectisol plant) to remove H_2S and reduce the CO_2 content.

TEMPERATURE		800 - 900 °C
PRESSURE		30 BAR
FEED MATERIAL		BEECHWOOD
ASH (DRY)		4.5 % BY WT
MOISTURE		15 % BY WT
GAS COMPOSITION		
H_2	% BY VOL	33.8
CO	% BY VOL	34.0
CO_2	% BY VOL	25.6
CH_4	% BY VOL	4.2
C_NH_M	% BY VOL	1.8
N_2	% BY VOL	0.6
H_2S	% BY PPM	10
SPEC. GAS PRODUCTION M^3_n/KG DAF		1,3
SPEC. OXYGEN CONSUMPTION M^3_n/KG DAF		0,26

TABLE XIV CFB Wood Gasification, Operating Pressure 30 bar, Production and Consumption Figures

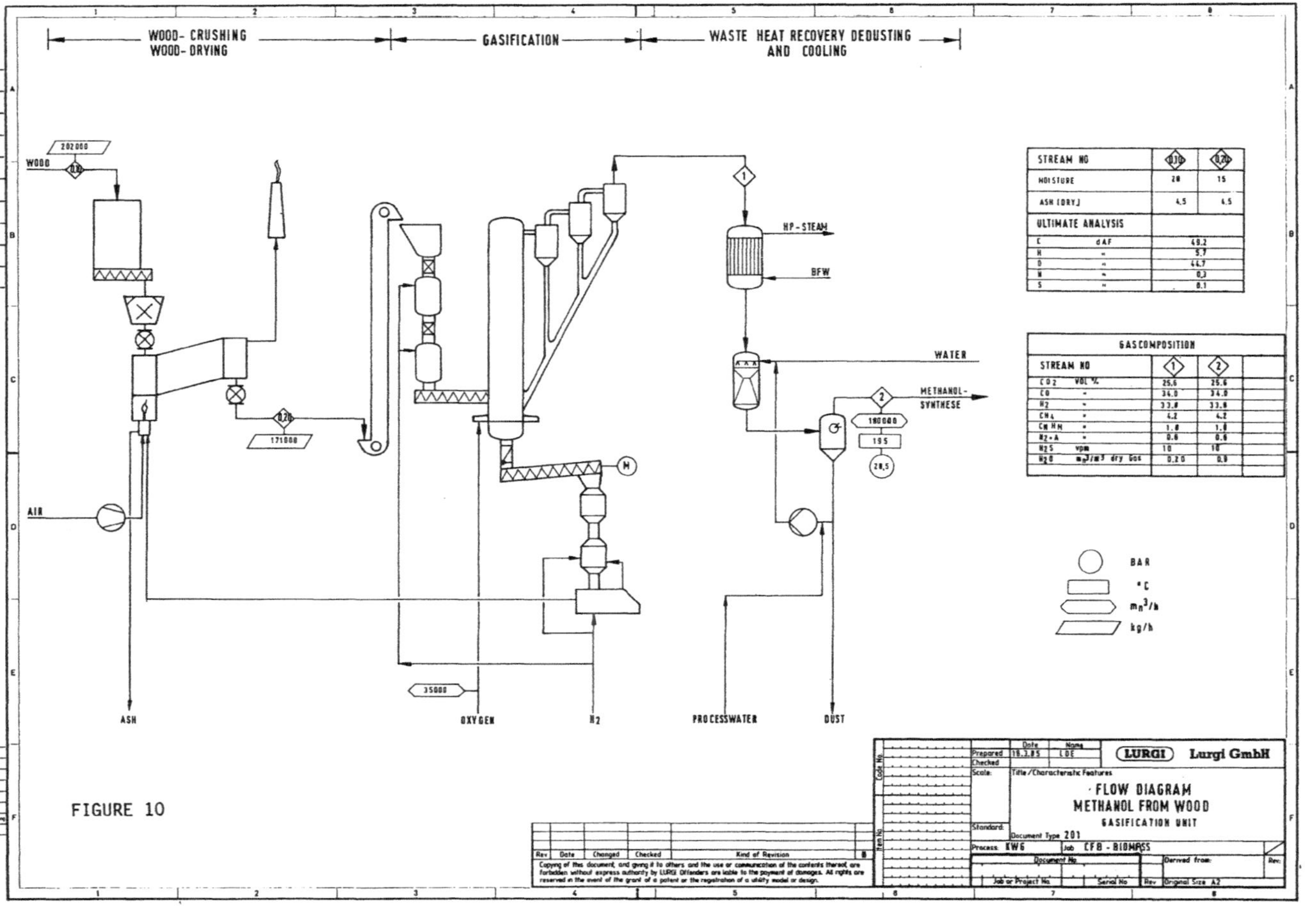

STREAM NO	0.10	0.20
MOISTURE	20	15
ASH (DRY)	4.5	4.5
ULTIMATE ANALYSIS		
C dAF	49.2	
H "	5.7	
O "	44.7	
N "	0.3	
S "	0.1	

GASCOMPOSITION			
STREAM NO	1	2	
CO_2 VOL %	25.6	25.6	
CO "	34.0	34.0	
H_2 "	33.0	33.0	
CH_4 "	4.2	4.2	
C_nH_m "	1.0	1.0	
N_2+A "	0.6	0.6	
H_2S vpm	10	10	
H_2O m_n^3/m^3 dry Gas	0.20	0.0	

FIGURE 10

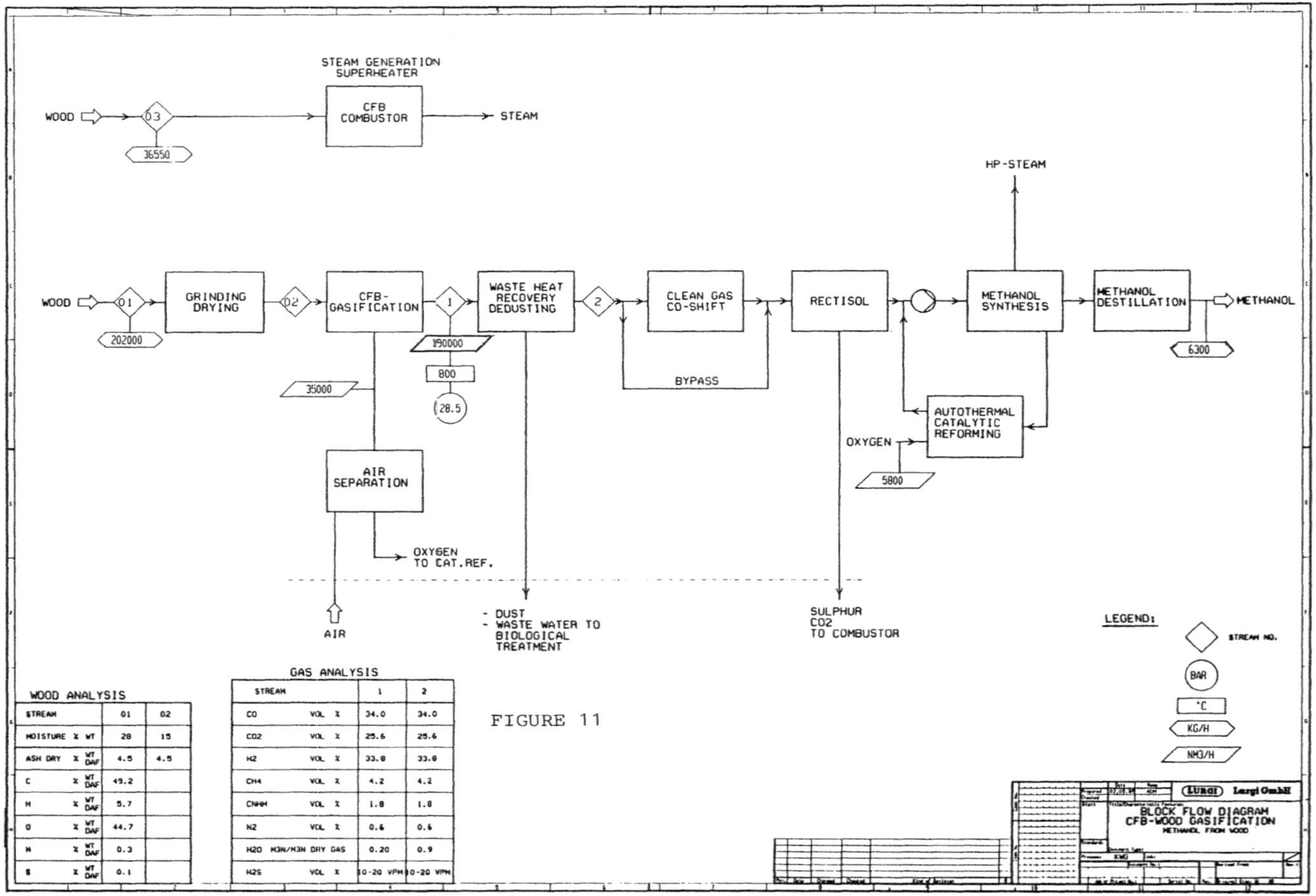

WOOD ANALYSIS

STREAM		01	02
MOISTURE	% WT	28	15
ASH DRY	% WT DAF	4.5	4.5
C	% WT DAF	49.2	
H	% WT DAF	5.7	
O	% WT DAF	44.7	
N	% WT DAF	0.3	
S	% WT DAF	0.1	

GAS ANALYSIS

STREAM		1	2
CO	VOL %	34.0	34.0
CO2	VOL %	25.6	25.6
H2	VOL %	33.8	33.8
CH4	VOL %	4.2	4.2
CNHM	VOL %	1.8	1.8
N2	VOL %	0.6	0.6
H2O	M3N/M3N DRY GAS	0.20	0.9
H2S	VOL %	10-20 VPM	10-20 VPM

FIGURE 11

The conditioned gas then enters the methanol synthesis unit. A detailed description of this process can be found in the relevant literature (16). In the case of CFB biomass gasification, the methane content of the feed gas requires the purge gas from the methanol synthesis loop to be fed to an authothermal catalytic reforming unit. The reformed gas, which consists only of CO_2, CO and H_2 (H_2O), is added to the stream from the Rectisol unit. This type of process configuration optimizes specific methanol production while oxygen consumption is minimized.

6.3 Overall Mass Balances and Thermal Efficiencies

The block flow diagram (Fig. 11) shows the production of chemical-grade methanol as well as the consumption figures. The specific figures for the plant can be taken from Table XV. The overall thermal efficiency based on wood consumption for gasification reaches 53 %, the specific methanol production amounts to 0,45 kg/kg DAF wood.
A comparison with the overall efficiency of 70 - 75 % for a methanol plant based on natural gas or naphtha shows the effect of the low HHV of wood parallel by a high oxygen content.

Considering that the dry wood consumption of the plant adds up to approx. 1.4 million tons/year, it becomes clear that energy farming is required for the production of the biomass source. Based on a realistic production figure of 10 t/h.y on forestry land, an area of about 1,200 km² is required.

	For gasification only	For gasification and steam 1)
Wood t DAF/h	138.5	163.5
Oxygen m^3n/h	40.000	40.000
Methanol t/h (chemical-grade)	63	63
Spec. Methanol Production kg/kg DAF	0.45	0.39
Spec. Oxygen Consumption kg/kg DAF	0.41	0.35
Thermal Efficiency %	53	45

1) Superheating and steam drive

TABLE XV Overall Production and Consumption Figures of "Methanol-from-Wood Plant"

6.4 Cost Considerations

A report on cost data must describe assumptions, inclusions and exclusions. The technological status of a process is one of the factors determining the accuracy of a cost estimate. In the case of methanol synthesis, all the downstream processes are commercially proven technologies, but a commercial-scale pressurized wood gasifier has not yet been operated. However, as this part of the process represents roughly 20 - 25 % of total investment costs, the uncertainty caused hereby is not severe.

The production costs of methanol from wood have been estimated for the plant as described above, making the following assumptions:

- self-supporting plant, but import of electrical energy
- Contingency 20 %
- Interest rate 10 % p. a.
- Depreciation time Case I 5 years
  Case II 17 years
- Escalation 4 % p. a.
- Duration of erection 3 years
- Salaries 1 % of TIC
- Maintenance 3 % of TIC
- Chemicals, process water 1 % of TIC
- Insurance and taxes 2 % of TIC

The method used was a dynamic annuity method for cost estimation. As a result, the annuity factors for dynamic capital costs (ACC) and the annuity factors for dynamic fixed operating cost (AOC) have been determined as follows:

		CASE I	CASE II
ACC	%	24,88	9,17
AOC	%	7,00	7,00

Figure 12, derived from these data, shows that at a realistic wood price of between 90 and 140 DM/dry ton, methanol prices range between 490 and 630 DM/t (taking CASE I, depreciation period 5 years). Comparing this figure with the present market price, the price of methanol-from-wood via CFB is clearly still roughly 30 % higher.

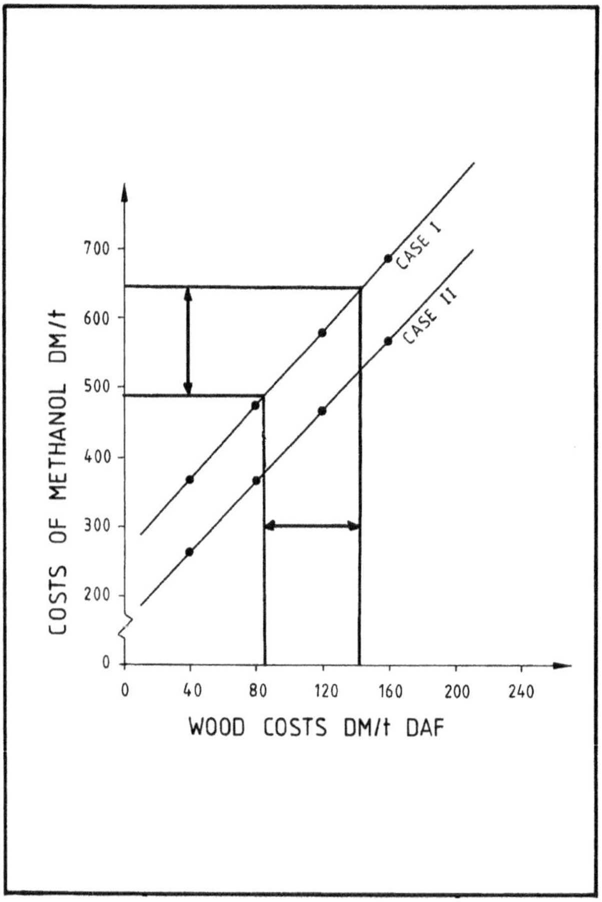

FIGURE 12 Methanol Costs in relation to on Wood Price for a 1,500 t/d Plant

7. Conclusion and outlook

The development reported herein has now been concluded and the atmospheric version of CFB biomass gasification is commercially available. Potential applications for syngas plants based on CFB biomass gasification are

- o production of chemical-grade and fuel methanol
- o production of fertilizer

A pressurized CFB gasifier has to be used to ensure that the requirements of these processes are fully met. In view of the facts reported here it becomes obvious that this pressurized gasifier needs pilot and demonstration units as further development steps.

However, a severe restriction becomes obvious when looking at the forestry area required for a medium-size methanol-from-wood plant. Therefore one has to look for potential applications not dealt with herein for smaller plant sizes. The most promising and economically attractive applications (14) are

- o production of fuel gas for direct heating of kilns in the paper, ceramic, and cement industries
- o power generation
- o combined power cycle (pressurized CFB gasification).

Further steps in that direction will

- help in refining the gasification technology
- ensure, that the technology is kept alive
- lower the financial commitments for the first commercial applications.

8. Literature

1) Reh, L.; Erdöl, Kohle, Erdgas, Petrochemie 32, 12, 560-566 (Dez. 1979)

2) Meunier, S.; Vergasung fester Brennstoffe, Verlag Chemie Weinheim (1962)

3) Hedden, K.; Wilhelm, A.; German Chemical Engineering 3, 142 ff, (1980)

4) van den Aarsen, F.G.; Beenackers, A.A. C.M.; van Swaaj, W.P.M.; Cocurrent Moving Bed and Fluidized Bed Biomass Gasification in (5)

5) Grassi, G.; Palz, W.; Editors: Energy from Biomass, Vol. 3, D. Reidel Pub. Comp. Dordrecht (1982)

6) Schrader, L.; Nitschke, E.; Will, H.; Bellin, A.; Biomasses as Feed for the Rheinbraun HTW-Gasification Process BGT-Conference on Energy Buena Vista, Fla., USA; 02.1984

7) Reimert, R.; ISPRA Course "Synthetic Fuels" May 7th - 11th, 1984 Commission of the European Communities, Joint Research Centre Ispra, Italy

8) Chrysostome, G.; Lemasle, J. M.; Pressurized Oxygen Blown Fluidized Bed Gasification of Wood in (9)

9) Palz, W.; Pirrvitz, D.; (Editors) Energy from Biomass, Vol. 5, D. Reidel Pub. Comp. Dordrecht (1983)

10) Mudge, L.K.; Becker, E.G.; Mitchel, D.H.; Catalytic Generation of Methanol Synthesis from Wood in (11)

11) Reed, T.B.; Graboski, N.; Editors;
Proc. Biomass-to-Methanol Specialists Workshop, Tamorron Durango, Colorado (1982)

12) Supp, E.; Lurgi's Improvements and Developments in Methanol Production and Conversion Technologies
AICHE Spring Meeting, Anaheim Ca, USA, May 1984

13) Reimert, R.; Lindner, Chr.; Mehrling, P.;
Bio Energy World Conf., Gothenburg 1984

14) Mehrling, P.; Reimert, R.; Fuel- and Synthesis Gas from Biomass via Gasification in the Circulating Fluid Bed
3rd EC Conference on Biomass, Venice, March 1985

15) "Apha" Standard methods for the Examination of Water and Waste Water (American health organisation) 14th Ed. 1975

16) Marschner, F.; Möller, F-M.; Gelbke, H.P.;
Methanol; Ullmann's Enzyclopaedia of Technical Chemistry, 4th Ed., Vol. 16 (621-633)

THE GASIFICATION OF WOOD USING THE OXYGEN DONOR PROCESS

Authors	:	I Dallas, R J McLellan, K E Scanlon, D H Smith

Contract No.	:	ESE/P/069/UK
Duration	:	24 Months 01.04.82 - 31.03.84 4 Months Extension 01.04.84 - 31.07.84
Total Budget	:	£848,423 C.E.C. Contribution : £339,334

Head of Project	:	Mr K E Scanlon, Wellman Mechanical Engineering Limited
Contractors	:	John Brown Engineers and Constructors Limited, 20 Eastbourne Terrace, London, W2 6LE
	:	Wellman Mechanical Engineering Limited, Roberts House, Smethwick, Warley, West Midlands

SUMMARY

The Oxygen Donor Gasifier produces a medium calorific value gas low in nitrogen by supplying the oxygen necessary for gasification from the reduction of calcium sulphate. The calcium sulphide formed is reoxidised to calcium sulphate using air in a separate compartment of the reactor and reused. The process utilises a fluidised bed to continuously transfer bed material containing the calcium sulphide/sulphate between oxidiser and gasifier compartments within the reactor. The heat required for the gasification process is generated in the oxidiser and supplied to the gasifier as sensible heat in the circulating fluidised bed.

A pilot plant designed to gasify 550 kg/hour of wood chip was built and commissioned. After overcoming numerous problems, mainly associated with feeding wood chip to the fluidised bed reactor, four trials were completed, the longest lasting ten hours with an average wood feed rate of 342 kg/hour. The plant consistently produced a gas free of tar and oil having a gross calorific value of 13.58 MJ/m³ (@ 0°C). Mass and energy balances have been derived from the experimental data and the results and their implications are discussed

CONTENTS

1.0 INTRODUCTION

The gasification of solid fuels to produce a gas for use as a fuel or chemical feedstock has been practised worldwide for many years.

The Oxygen Donor Gasification (ODG) process offers the potential advantages of oxygen gasification (and thus a higher calorific value product gas) without the need for an independent oxygen supply.

The ODG concept evolved as a result of work done by Exxon in the UK on chemically active fluid beds. The process is being developed by a Joint Venture between Wellman and John Brown.

1.1 THE ODG CONCEPT

Oxygen Donor Gasification (ODG) is a process where the oxygen necessary for gasification is supplied by the reduction of calcium sulphate to calcium sulphide, the basic gasification reactions being:-

$$CaSO_4 + 4C \rightarrow CaS + 4\,CO$$

$$CaSO_4 + 2C \rightarrow CaS + 2\,CO_2$$

Intermediate reactions also occur including :-

$$H_2O + C \rightarrow H_2 + CO$$

$$CaSO_4 + 4H_2 \rightarrow CaS + 4H_2O$$

$$2H_2O + C \rightarrow 2H_2 + CO_2$$

$$CO_2 + C \rightarrow 2CO$$

The calcium sulphide generated is oxidised using air as the source of oxygen thus reforming calcium sulphate for further use in the gasification process, the reaction being:-

$$CaS + 2O_2 \text{ (From Air)} \rightarrow CaSO_4$$

The above reactions take place in a fluidised bed reactor containing two separate compartments, a gasifier and an oxidiser. The bed material (See para 2.1.3) which contains the calcium sulphate/sulphide reactant is circulated between the two compartments thus providing a continuous source of oxygen for gasification.

The gasification reactions are heavily endothermic, the gasifier temperature is maintained by supplying it with hot bed material from the oxidiser. The reoxidation of calcium sulphide being exothermic supplies part of the energy required to heat the bed material in the oxidiser, the shortfall being supplied by the burning of carbon carried from the gasifier by the circulating bed material. Both the gasification and oxidation reactions commence at 850°C and the reactor has been designed to operate with the oxidiser at 1000°C and the gasifier at 950°C.

For operation at ambient pressures and the above temperatures the predicted gas composition was (on a dry basis) :-

Component	Volume %
CO	36
CO_2	23
H_2	40
CH_4	1

In practice, the gas obtained during the pilot plant trials contained more methane and generally had a composition (on a dry basis) :-

Component	Volume %
CO	37.5
CO_2	14.5
H_2	31
CH_4	12.2
N_2	4.8

1.2 ADVANTAGES OF THE ODG PROCESS

The advantages of the ODG process over other gasification systems lie in its ability to produce a medium calorific value gas, low in nitrogen, without the use of pure oxygen and hence expensive cryogenic plant. In addition, by introducing fuel into a high temperature fluidised bed it is rapidly pyrolysed with the products of pyrolysis held at a high temperature for a relatively long time thus cracking hydrocarbons and reducing the quantity of troublesome tars in the product gas.

In summary, the potential advantages of the ODG process are as follows:-

- It produces a medium calorific value gas typically with a gross CV of 12 - 14 MJ/Nm³ (300 - 350 Btu/scf) without the use of pure oxygen.

- It produces a gas low in nitrogen.

- The gas produced is relatively clean because of the high operating temperature (950 - 1000°C).

- The system is tolerant to a wide range of feedstocks.

- It produces a gas of low sulphur content from high sulphur feedstocks due to the ability of the bed material, which contains calcium oxide, to absorb sulphur.

- It starts up rapidly and responds readily to changes in demand.

1.3 AIM OF THE PROJECT

This was to construct a Pilot Plant capable of gasifying 550 kg/h of wood and to carry out the following experimental work :-

- to establish a procedure for starting up the pilot plant
- to demonstrate that the plant and process will operate under stable conditions
- to demonstrate the principles of the ODG process
- to observe the behaviour of the dual-bed stone recirculation system and the stone material
- to demonstrate the efficiency with which C in the wood is converted to C in the CO of the product gas
- to determine the efficiency with which the calorific value of the wood is converted to calorific value in the product gas plus recoverable heat in product and flue gases
- to gasify wood at the rate of 12 tonnes per day and produce a medium calorific value gas suitable for conversion to methanol

It was not our objective, within the programme, to optimise the ODG process or its economics.

2.0 DESIGN OF THE PILOT PLANT

The major elements of the ODG process are indicated in the process flow diagram (PFD) figure 1 and the process flows, temperatures and pressures are given in table 1.

2.1 DESIGN BASIS

2.1.1 Wood Feed Composition and Properties

The PFD was based on mass and energy balances produced for a system capable of gasifying 550 kg/h of wood with the composition :-

Ultimate Analysis:-	Carbon	48.76 wt%
	Hydrogen	6.00 wt%
	Oxygen	44.18 wt%
	Nitrogen	0.49 wt%
	Sulphur	0.01 wt%
	Ash	0.56 wt%

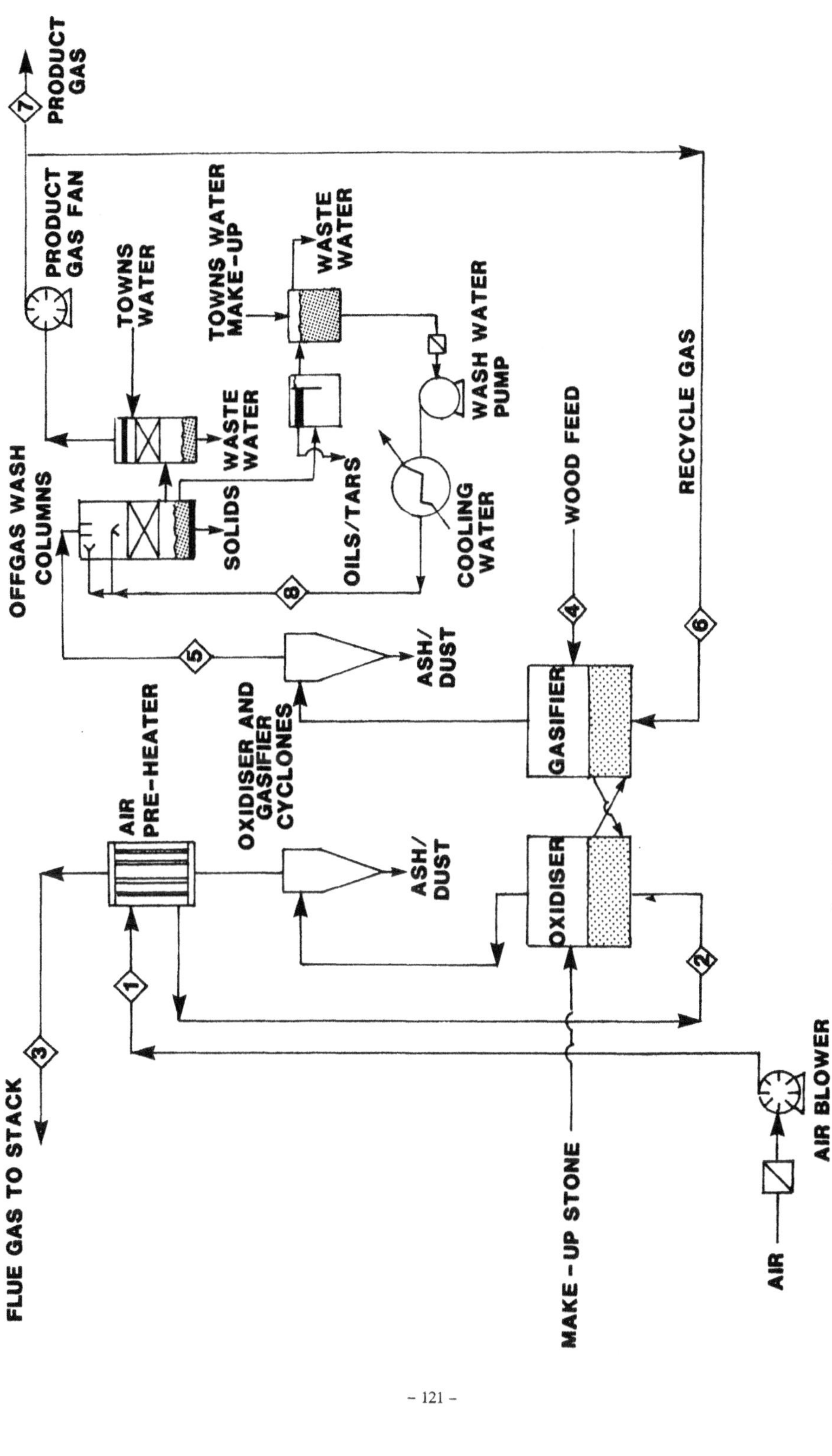

ODG - PROCESS FLOW SCHEME

FIGURE 1

DESIGN MASS FLOWS, OPERATING TEMPERATURES & PRESSURES

NUMBER	1	2	3	4	5	6	7	8
DESCRIPTION	FLUID-ISING AIR	FLUID-ISING AIR	FLUE GAS	WOOD FEED	OFF GAS	RE-CYCLE GAS	PRO-DUCT GAS	WASH WATER
MASS FLOWS kg/h	1627	1627	1562	550	1328	713	423	26178
OPERATING TEMP°C	25	500	553	25	950	20-90	20–90	35
OPERATING PRESS mBAR	1331	1305	1035	1013	1068	1317	1013	4470

TABLE 1

Moisture content = 20 wt%

Gross calorific value :- 18,590 kJ/kg dry ash free basis.

Size :- Chips with a max size of 1cm x 0.5cm x 0.5cm

2.1.2 Wood Gas Composition

The predicted gas composition, based on data from the modified Winkler process, was:-

	wt%	Mole %
Carbon Monoxide	42.17	31.52
Carbon Dioxide	43.58	20.67
Hydrogen	3.36	35.07
Methane	0.41	0.63
Water Vapour	10.48	12.11

The modified Winkler process uses an oxygen and steam mixture for gasification at 975°C. Gas compositions from other gasifiers have higher methane contents, but the above composition was chosen for design purposes as it requires the largest heat for reaction and produces the largest gas volume.

2.1.3 Choice of Bed Material

Criteria for the bed material were :-

- it should contain sufficient calcium sulphate to satisfy the oxygen donation reaction
- it should contain calcium oxide capable of absorbing sulphur containing compounds from the gasification process, the typical reaction being:-

$$CaO + H_2S \rightarrow CaS + H_2O$$

- the material should be mechanically strong so as to ensure a satisfactory life under the erosive operating conditions
- the particle size should be 0.8mm dia ± 0.2mm
- it should be readily available and inexpensive

Of the materials investigated, only limestone and an Agglomerated Portland Cement were mechanically strong enough to withstand thermal cycling and fluidisation for a reasonable time span. There were problems in obtaining both materials in the size range required and due to its availability and cheapness limestone of various sizes was used throughout the trial.

The most satisfactory material tested was a limestone with a size distribution:-

0.76mm - 0.57mm	70 wt%
<0.57mm	30 wt%

Before gasification, the limestone was calcined for several hours in the oxidiser and then sulphated with sulphur dioxide fed into the oxidiser air supply. In this way a material consisting essentially of calcium oxide but containing about 5 percent of calcium sulphate was obtained.

2.1.4 Assumptions Used in Designing the Reactor

The major assumptions used in the design of the reactor were:-

- the gasifier is fluidised using recycled gas at essentially ambient temperature
- the recycled gas reaches gasifier operating temperature instantaneously on contact with the hot fluidised bed
- the recycled gas composition does not change on passing through the gasifier
- the recycled gas flow is only that needed to provide incipient fluidisation at the operating temperature of the bed.
- all volatile matter in the fuel is converted to low molecular weight gaseous hydrocarbons without the formation of tar
- fuel residue forms as char in the bed, some of which transfers with the bed material to the oxidiser where it burns to generate heat in the system
- all sulphur present in the fuel forms calcium sulphide
- the oxidiser is operated with 10% excess air
- the oxidiser is fluidised using air preheated to 500°C and the air reaches the oxidiser operating temperature instantaneously on contact with the hot bed material
- bed material of 0.8mm mean diameter is used and this determines the minimum fluidising velocity required

2.2 PREDICTED MASS AND ENERGY BALANCES

The following mass and energy balances relate only to the gasifier/oxidiser reactor and its associated dust cyclones.

2.2.1 Predicted Mass Balance

INPUT	kg/h	OUTPUT	kg/h
Wood	550	Gas after Cyclone	1328
Air	1627	Flue Gas	1562
Make-up Bed Material	53	Gasifier Cyclone Fines	20
Recycled Gas	713	Oxidiser Cyclone Fines	33
TOTAL	2943		2943

2.2.2 Predicted Energy Balance

INPUT		GJ/h	OUTPUT		GJ/h
Wood	CV	7.383	Gas from Cyclone	CV	10.523
Preheated Air	SH	0.718		SH	1.935
Recycled Gas	CV	6.611		LH	0.438
	SH	0.029	Flue Gas to Preheater	SH	1.429
				LH	0.025
			Gasifier Cyclone Fines	CV	0.070
				SH	0.015
			Oxidiser Cyclone Fines	SH	0.025
			Heat Losses		0.281
TOTAL		14.741	TOTAL		14.741

CV = Calorific Value (Chemical Energy)
SH = Sensible Heat
LH = Latent Heat
Basis: 25°C, water as liquid

2.2.3 Heat and Mass Transfer between Oxidiser and Gasifier

From mass and energy balance calculations, the bed material is required to transfer oxygen to the gasifier at a rate of 142 kg/h and heat at a rate of 3.43 GJ/h. Based on the assumed temperature difference between the gasifier and oxidiser of 50°C the quantity of bed material which has to be transfered to satisfy the heat balance is 57,000 kg/h.

Based on the above transfer rate, a bed material containing 0.5 wt% calcium sulphate would be sufficient to supply the oxygen required for gasification of wood fed at a rate of 550 kg/h.

2.3 BED TRANSFER MECHANISM

Successful operation of the ODG process depends on the continuous transfer of large quantities of hot bed material between oxidiser and gasifier whilst maintaining independent product and flue gas streams. To reduce heat losses, the gasifier and oxidiser were incorporated into a single reactor vessel and the method used to circulate bed material between the two compartments was as shown in Figure 2. The transfer system consists of two zones which differ from the normal fluidised bed, these are a transfer slope and a pump section. Transfer of bed material from one compartment to another is caused by material directly above the transfer slope slumping onto it and being directed into the pump section of the adjacent compartment where vigorous fluidisation lifts the fresh bed material into the main fluidised bed. As material is being removed from one side of the fluidised bed and added to the other a differential hydrostatic head is created which acts as a driving force to circulate the bed.

To maintain the oxidiser and gasifier as separate vessels, two transfer systems are required, one transferring fresh material from the oxidiser to the gasifier and the other transferring used material from the gasifier back to the oxidiser.

2.4 REACTOR SHAPE

The fluidised bed areas required were calculated from the predicted mass balance and a maximum free-board gas velocity of 2 m/s.
With an additional allowance for operating flexibility the areas chosen were :-

Oxidiser	1.6 m^2
Gasifier	1.0 m^2

In order to achieve the high bed circulation rate required (57 tonnes/h) it was logical to make the transfer sections of the bed as large as possible without restricting flows in other areas of the reactor or having areas of bed which were easily bypassed. Numerous designs were investigated and the layout finally chosen is illustrated in Figure 3. The design not only satisfied the above criteria, but was also easy to construct.

2.5 THE COLD MODEL

As no design data was available on the chosen mechanism for bed transfer it was necessary to generate this experimentally. To this end a full size model of the gasifier/oxidiser reactor was built and the effect of the following variables on bed transfer rate were investigated :-

- the depth of material in the fluidised bed.
- the area of the transfer slot
- the ratio of fluidising velocities in the main area (fluidised section) and pump section of the bed
- the effect of fluidising the material above the transfer slope

DETAIL OF BED TRANSFER MECHANISM

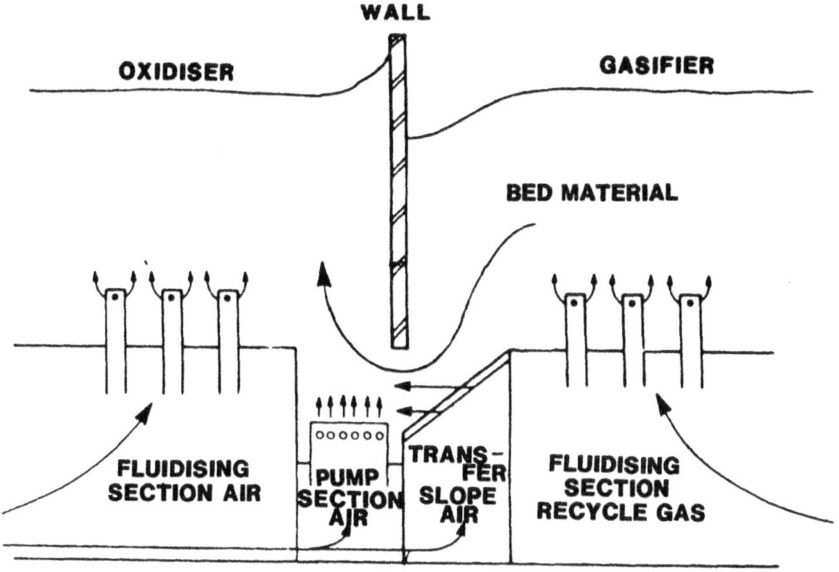

FIGURE 2

REACTOR SHAPE

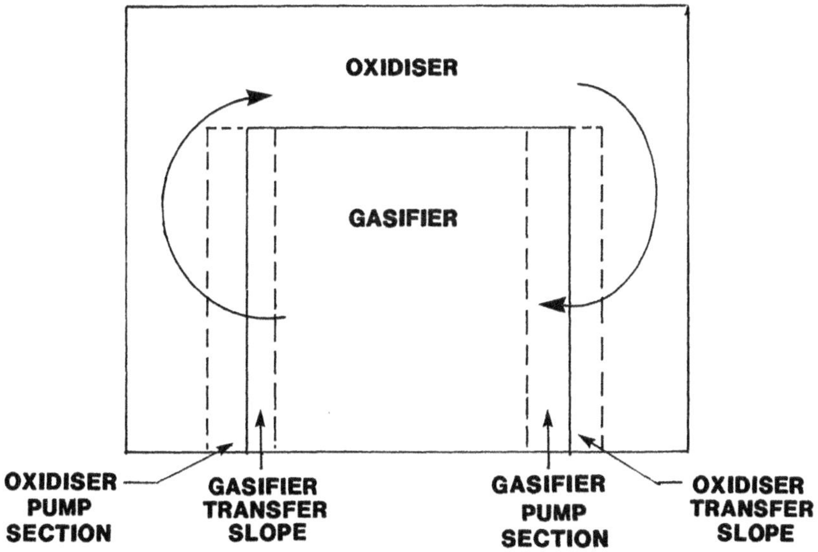

FIGURE 3

Due to time limitations, the tests were only used to produce data which proved that a bed transfer rate of 57 tonnes/hour could be achieved. Transfer section geometry and operating conditions were not optimised.

The Cold Model proved invaluable not only in supplying the needed design data but also in providing general experience of circulating fluidised bed behaviour.

The bed material used in the Cold Model tests was sand with an average particle size of 0.5 mm diameter and cold air was used as the fluidising medium.

2.5.1 Determination of Bed Transfer Rate

In order to produce data from the Cold Model a reliable method of measuring the bed circulation rate had to be established.

A float on the surface of the bed was used as a rough indication of the speed at which the bed travelled but its movement was often erratic and it was doubtful that both the bed material and float travelled at the same speed.

The method selected to determine the bed circulation rate involved the addition of a tracer to the gasifier compartment of the reactor and monitoring the rate at which the concentration of tracer in the gasifier and oxidiser compartments changed when the system was in operation. Potassium permanganate was the tracer used and its concentration was determined by leaching it from a fixed quantity of bed material and analysing the resulting solution colorimetrically.

Three sampling points were used in the Cold Model from which bed material could be collected, these were positioned as shown in Figure 4.

To measure the circulation rate, a quantity of tracer was added to the gasifier compartment and thoroughly mixed with the bed material by fluidising only the gasifier. The gasifier bed was then slumped and the valves controlling the air flow to the gasifier were adjusted to give the test conditions. The gasifier and oxidiser were then fluidised simultaneously and samples of bed material were collected at regular intervals from the three sampling points. Analysis of the samples for tracer produced concentration verse time plots of the form shown in Figure 5.

Analysis of tracer concentration verse time plots indicated that the combined reactor was best modelled with the gasifier as a continuous stirred tank reactor and the oxidiser as a plug flow reactor. This can be

LOCATION OF COLD MODEL BED SAMPLING POINTS

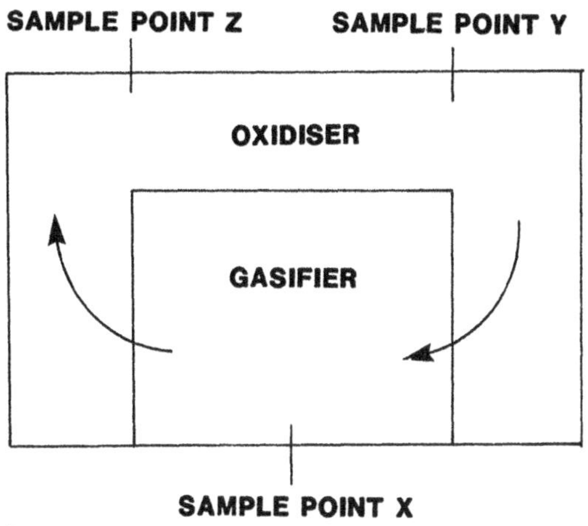

FIGURE 4

COLD MODEL TRACER CONCENTRATION VERSES TIME PLOTS

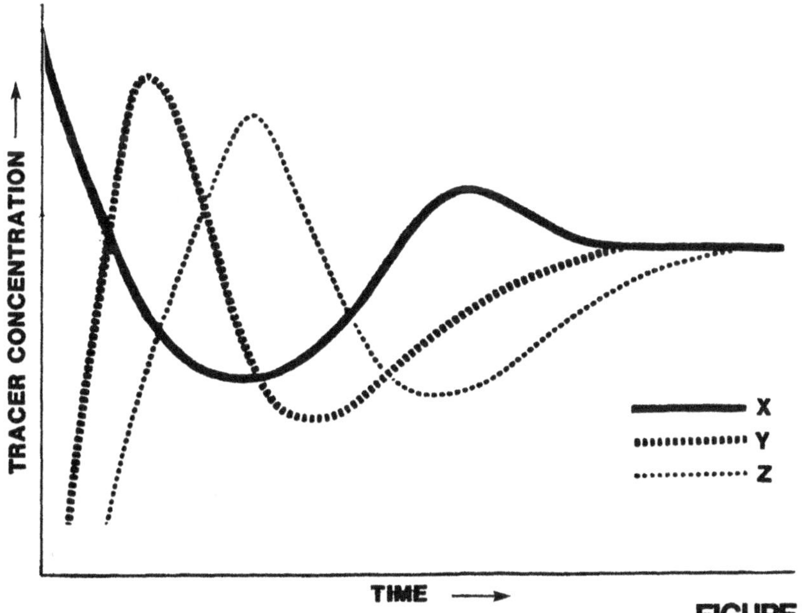

FIGURE 5

attributed to the relative shapes of the gasifier and oxidiser compartments, the gasifier being of a similar length and width while the oxidiser is much longer than it is wide.

Using the above model, three dimensions on the tracer verse time plot were used to deduce the actual circulation rate from the results obtained experimentally. These were :-

- the initial rate of decline in the tracer concentration in the gasifier.
- the time taken for the tracer to complete one cycle and return to the gasifier.
- the time taken for the tracer to travel from point Z to point Y in the oxidiser.

In each experiment the rates of bed circulation, calculated by the three methods were consistent.

2.5.2 Results of Cold Model Tests

(i) Effect of fluidised bed depth on circulation rate:-

Fixed operating parameters for the duration of the experiment were :-

- Transfer slot dimensions, width = 0.8m
  height = 0.113m
- Fluidising Velocity in Pump Section/Fluidising Velocity in Fluidising Section $(V_p/V_f) = 4$
- No fluidising air above transfer slope.

Slumped Bed Depth (metres)	Average Sand Circulation Rate (tonnes/hour)
0.155	21.8
0.265	36.3
0.265	34.8
0.310	38.8

The results plotted in figure 6 indicate that circulation rate increases proportionally with increasing bed height. The driving force for the movement of material through the transfer slot is the difference between the mass per unit area of material on either side of the slot. Thus, because of the difference in bed density resulting from different fluidising velocities on either side of the transfer slot, the higher the bed, the greater is the driving force.

EFFECT OF BED HEIGHT ON BED CIRCULATION RATE

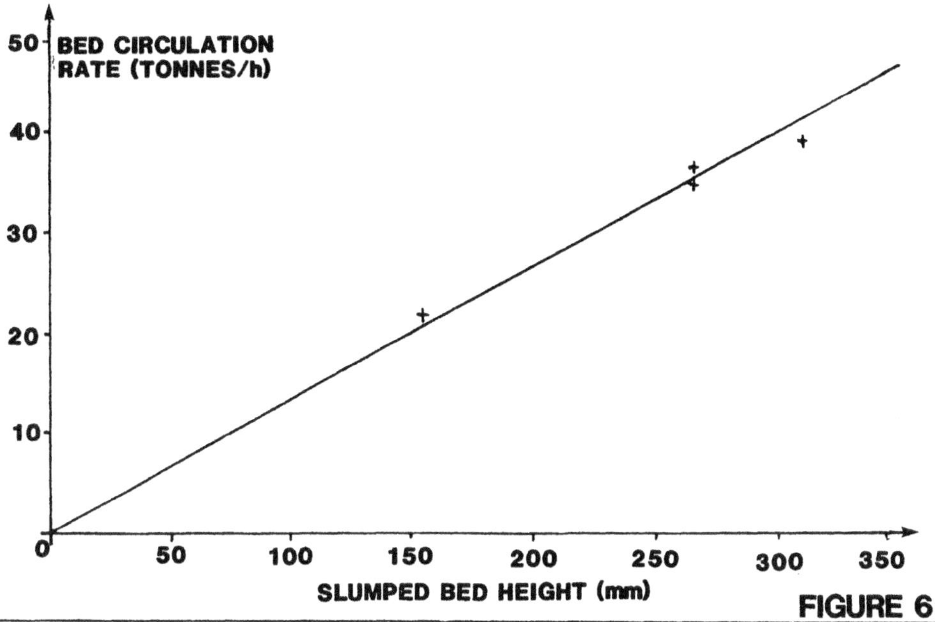

FIGURE 6

EFFECT OF Vp/Vf ON BED CIRCULATION RATE

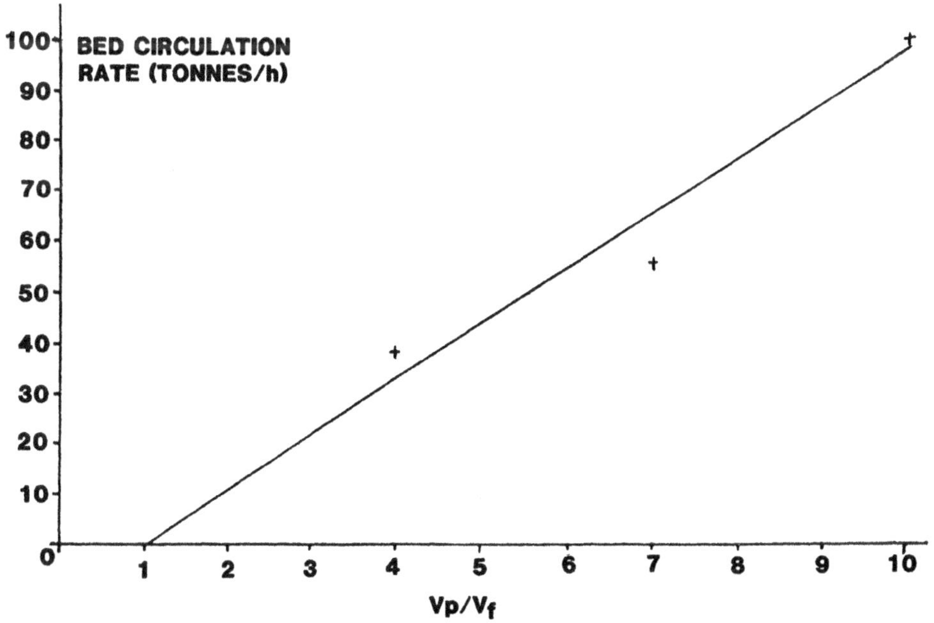

FIGURE 7

(ii) Effect of Transfer Slot Height on Circulation Rate.

Fixed operating parameters for the duration of the experiment were :-

- Bed depth = 0.265m

- $V_p/V_f = 4$

- No fluidising air above transfer slope.

The bed circulation rate was measured for minimum and maximum effective transfer slot heights, the results being:-

Transfer Slot Height (metres)	Average Sand Circulation Rate (tonnes/hour)
0.065	27.6
0.113	35.5

(iii) Effect of V_p/V_f on Circulation Rate.

Fixed operating parameters for the duration of the experiment were :-

- Transfer Slot Dimensions, width = 0.8m
  height = 0.113m

- Bed depth = 0.3m

- No fluidising air above transfer slope.

V_p/V_f	Average Sand Circulation Rate (tonnes/hour)
1	LOW
4	38
7	56
10	100

The results are plotted in figure 7 and again indicate that circulation rate increases proportionally with increasing difference in the ratio of fluidising velocity in the Pump and Fluidising sections of the bed.

(iv) Effect of fluidising air above transfer slope on circulation rate.

Three different conditions were investigated:-

- no fluidising air above transfer slope

- low flow of air through the transfer slope in order to aerate the bed material and increase its fluidity. The air flow was never sufficient to fluidise the bed above the transfer slope.

- high velocity horizontal air jets at base of transfer slope designed to direct bed material through the transfer slot and into the adjacent pump section.

The bed circulation rate was found to be independant of how the transfer slope was operated and the hypothesis that a slumped bed on the transfer slope would reduce the circulation rate was unfounded. It is possible that vibrations in the test rig influenced slumped bed stability more than aeration or jetting in the region of the transfer slope.

2.5.3 Overall Conclusions from Cold Model Tests

In the Cold Model bed circulation rates in excess of 57 tonnes/hour could be achieved.

The rate of bed circulation was readily controlled by altering the ratio of gas velocity in the pump sections and fluidising sections of the bed.

In order to avoid substantial ellutriation of bed material, the free-board needed to be larger than the 0.6m used in the Cold Model.

2.6 REACTOR DESIGN

The oxidiser and gasifier areas were already established (see section 2.4). The height of the reactor was based on a bed depth of 0.6m and a free-board of 1.5m giving a total inside height of 2.1m. Due to the increase in free-board the reactor was 1m higher than originally anticipated and this presented constructional problems as the use of a refractory wall separating gasifier and oxidiser compartments was no longer practicable and a metallic dividing wall was chosen in preference. This resulted in a problem of differential expansion between oxidiser refractory walls and metallic gasifier shell which was overcome by locating the gasifier compartment centrally in the oxidiser as shown in figure 8. The gasifier outlet duct was taken through the oxidiser casing and expansion bellows between the two provided a gas tight seal and compensated for movement due to differential expansion.

The oxidiser shell was of mild steel construction and refractory lined with 150mm of dense castable refractory and 150mm of insulating castable refractory.

The metallic gasifier shell was constructed entirely from 5mm thick Incoloy 800H material ribbed and braced to increase its

DESIGN FEATURES OF THE ODG REACTOR

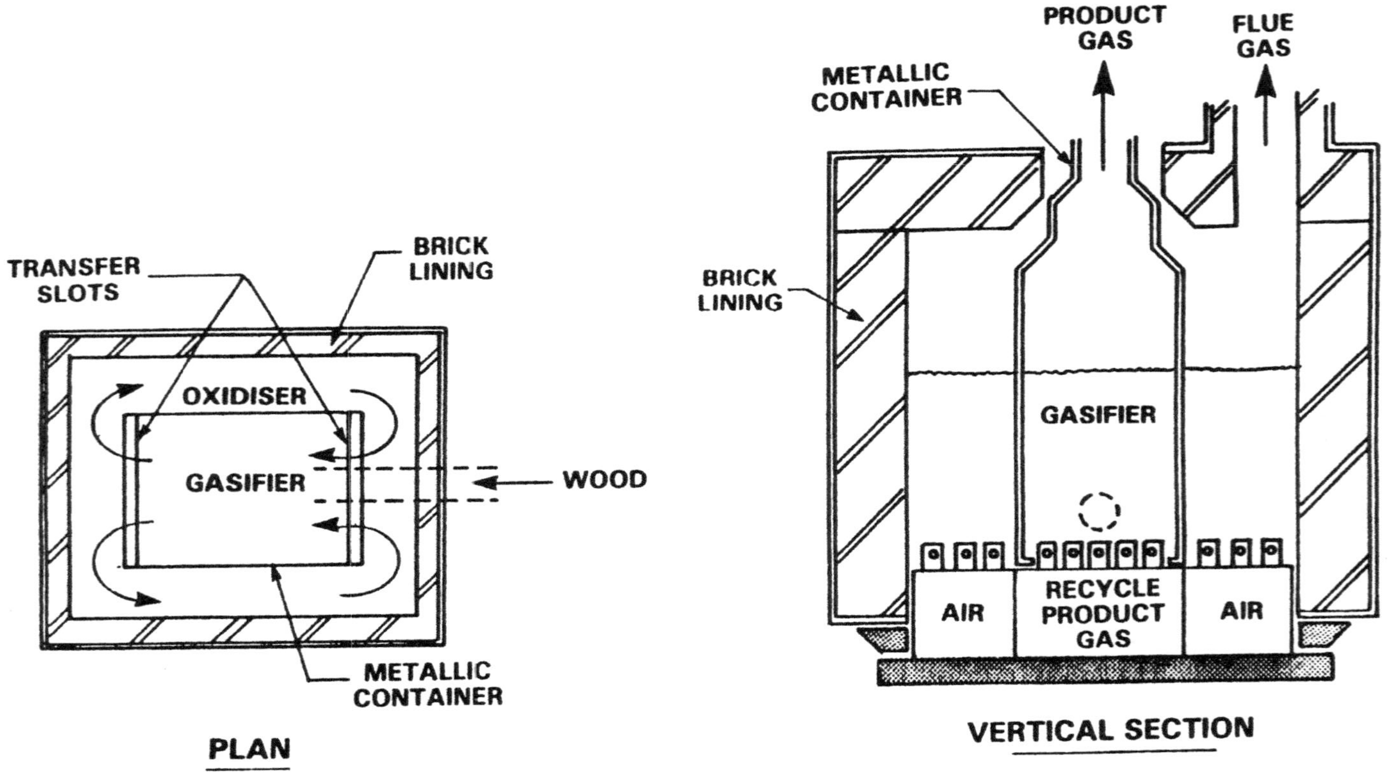

FIGURE 8

strength. Incoloy 800H is an austenitic solid solution alloy containing 33% Nickel and 21% Chromium.

Recycled gas and air for fluidisation entered the reactor through plenum chambers and nozzle type distributors. To encourage a uniform pressure within the oxidiser plenum the air was supplied to several locations from a common manifold. The plenum chamber and distributor nozzles were fabricated from 310 stainless steel, (25% Cr, 20% Ni).

2.6.1 Start-up System

Two methods of heating the reactor to its operating temperature were considered, these being:

- The combustion of a natural gas/air mixture supplied to the bed through the distributor nozzles. Initially the gas would burn above the surface of the bed material but as the bed heated the flame front would travel into the bed and heat it to the operating temperature.

- The combustion of natural gas over the surface of the oxidiser bed introduced at side burners mounted in the oxidiser walls. In this case the bed material would only attain the temperature of the pre-heated fluidising air and in order to increase it to the operating temperature solid fuel would have to be burnt within the bed.

The second of the above two methods was chosen owing to the risk of an explosion within the plenum chamber caused by the formation of air/gas mixtures.

Four burners were installed in the oxidiser casing, each capable of supplying 0.5GJ/hour and having a turndown of 20:1. The burners were supplied with air and gas from common control valves. Trimming of individual burners was performed manually by control valves mounted on the burner head. Incorporated in each burner was a flame failure probe. The burner control system ensured that if any burner failed the gas supply to all four was shut off. A small centrifugal type gas booster was used to supply the natural gas at a pressure of 1150 mbar (abs). Combustion air was taken from the fluidising air line on the air blower discharge. In order to prevent bed material blocking the burner ports the combustion air was maintained at a low flow even when the system had reached operating temperature and the burners had been turned off.

Once the bed reached 500°C it was planned to introduce a solid fuel (wood or coal) into the Oxidiser through the make-up stone feed system and thus increase the bed temperature to is operating condition.

2.6.2 Safety Considerations

Explosions resulting from the formation of product gas/air mixtures were considered to be impossible because the temperature at which gasification occured was sufficient to ensure that any product gas which came into contact with air would immediately burn.

For protection against the unlikely event of a natural gas/air explosion, two 0.46m x 0.69m explosion panels were incorporated in the oxidiser casing. To prevent the pressure in the oxidiser and associated equipment from rising above the 1550 mbar (abs) design value, the explosion panels needed to fail at 1130 mbar (abs). Proprietary bursting panels were unavailable and numerous materials were tested for gas permeability and bursting pressure, magnitude and repeatability. The material found most suitable was 0.25 mm thick Klingerit (a compressed asbestos fibre) and panels of this material were used throughout the trials. A 0.3m thickness of loosely packed ceramic fibre protected the panels from high temperatures within the oxidiser.

A 0.15m diameter graphite bursting disc was used to protect the gasifier and associated equipment against overpressurisation.

During the final process and mechanical design the system was subjected to a detailed hazard and operability study. This high-lighted a number of minor problem areas and the design was modified to eliminate them.

2.6.3 Make-up Stone Feed System

The items comprising the stone feed system are shown in figure 9. With the hopper outlet valve closed material was manually loaded into the 0.6m³ hopper via a coarse screen. The upper valve was then closed and the lower one opened. A vibrating bin discharger fed material from the hopper to the inlet of a 0.15m diameter variable speed screw conveyor which transferred the material to the oxidiser fluidised bed. A capacitance type level switch was mounted near the base of the hopper to detect a low level.

2.7 FUEL HANDLING AND FEEDING SYSTEM

The fuel handling system, shown in figure 10, was designed to convey woodchip (sized 1cm x 0.5cm x 0.5cm) at a maximum rate of 0.5 tonne per hour from a fuel reception area into the bottom of the gasifier fluidised bed.

Fuel was loaded into a reception bunker using a Front End Loader. A belt type discharge conveyor, fitted in the base of the bunker,

MAKE-UP STONE FEED SYSTEM

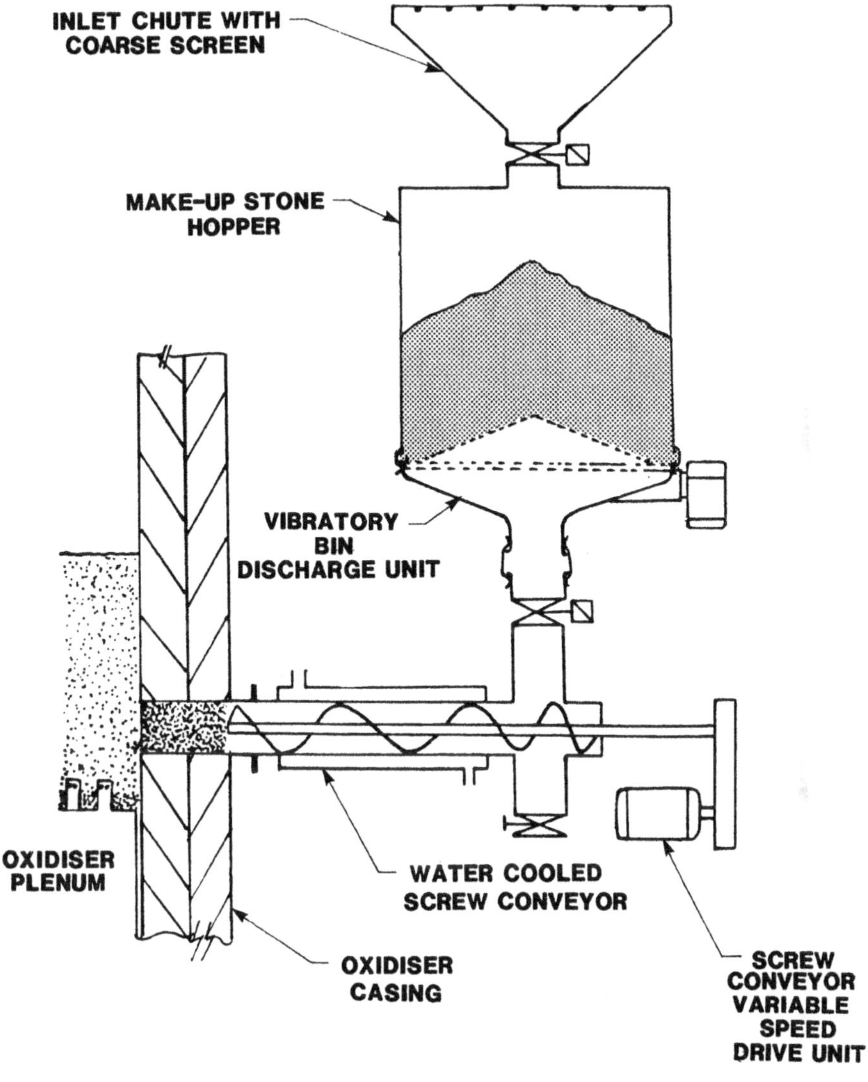

FIGURE 9

FUEL HANDLING SYSTEM

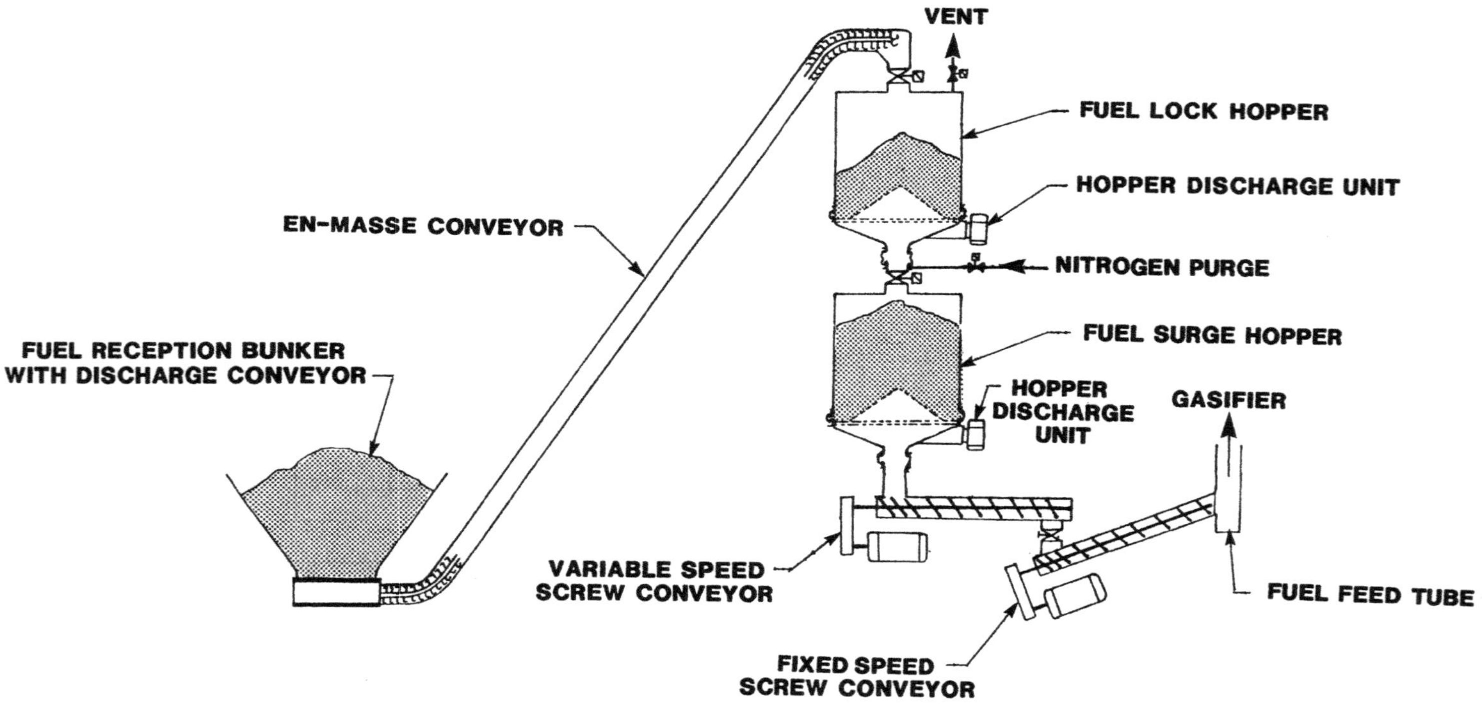

FIGURE 10

fed fuel to the inlet of an En-masse chain conveyor which in turn transported it to the inlet of a double lock hopper system. Each of the hoppers, one mounted above the other, was fitted with a vibratory bin discharge unit which allowed free flow of woodchip. A horizontal variable speed screw conveyor received fuel from the bottom hopper and fed it at a fixed rate to a fast speed inclined screw conveyor. The inclined conveyor deposited the fuel at the base of the gasifier fuel feed tube. Recycled gas passed into the feed tube through a sieve plate to transport the fuel into the gasifier bed and at the same time fluidise the bed material in the feed tube.

2.7.1 Fuel Handling

i) Reception Bunker

The mild steel bunker was situated at ground level and had a capacity of approximately 2.5 tonne of woodchip. Incorporated in the base of the bunker was a full length heavy duty rubber conveyor belt with transverse cleats to assist movement of the fuel. An adjustable feed gate on the bunker outlet controlled the rate at which fuel was taken from the unit.

ii) 'En-Masse' Chain Conveyor

The mild steel 'En-masse' conveyor was specially designed to handle woodchip. It had a head mounted drive unit and was inclined to a height of 10m above the floor. Fuel leaving the conveyor passed through a 200mm diameter slide valve into the upper fuel hopper.

iii) Fuel Feed Lock Hoppers

The top feed hopper was designed to hold 0.5 tonne of woodchip and comprised a gas tight mild steel cylindrical hopper 1.2m diameter x 1.8m high with conical base and sealed at the inlet and outlet with knife type slide valves. Situated between the conical base and main body was a 1.2m diameter bin discharge unit fitted with an externally mounted vibrating motor. A pressure tight flexible sleeve joined the vibrating cone base to the static hopper.

Connected to the hopper were nitrogen purge and vent lines, these were used to purge the hopper of gas prior to filling with wood and then to purge the air from the wood before feeding it to the lower hopper.

High and low level switches of capacitance type

were set in the wall of the hopper to indicate maximum and minimum levels of fuel.

Fuel discharging from the upper hopper fell into the lower hopper, which was of identical construction and acted as a surge hopper for the variable speed screw feeder.

A programmable logic controller (PLC) was used to control fuel level in the hoppers as well as the nitrogen purging sequence.

2.7.2 Fuel Feeding to Gasifier

i) Variable Speed Screw Conveyor

The woodchip leaving the bottom hopper passed through a chute into a 150mm diameter screw conveyor which was driven by a 2.2 KW motor via a magnetic drive variable speed controller. The screw conveyor was mounted horizontally and of mild steel construction. Double pitch blading at the start of the screw, followed by single pitch, ensured that the screw only operated half full of material. Fuel leaving the conveyor passed down a vertical tube containing a manual isolating valve and into the second screw conveyor.

The variable speed controller enabled fuel to be fed to the second screw feeder at any constant rate up to 0.5 tonnes/h.

ii) Constant Speed Screw Conveyor

This was a 150mm diameter mild steel screw conveyor driven by a 2.2 KW geared motor unit. The inclined screw was cantilevered off bearings at its inlet, leaving the discharge end open to ensure an unobstructed flow of fuel into the gasifier fuel feed tube. To protect the screw conveyor from excessive temperatures its end terminated 150mm short of the gasifier feed tube.

iii) Gasifier Fuel Feed Tube

This was constructed in type 310 stainless steel as indicated in figure 11. Material in the 150mm diameter tube was fluidised by recycle gas which passed through a sieve type distributor plate. Fuel deposited at the base of the tube was expected to mix with the bed material and be transported into the main gasifier bed where gasification would occur.

2.8 PRODUCT GAS CLEAN-UP AND BOOSTING SYSTEM

The hot gas leaving the gasifier was passed through a cyclone to

DETAILS OF Mk.1 FUEL FEED TUBE

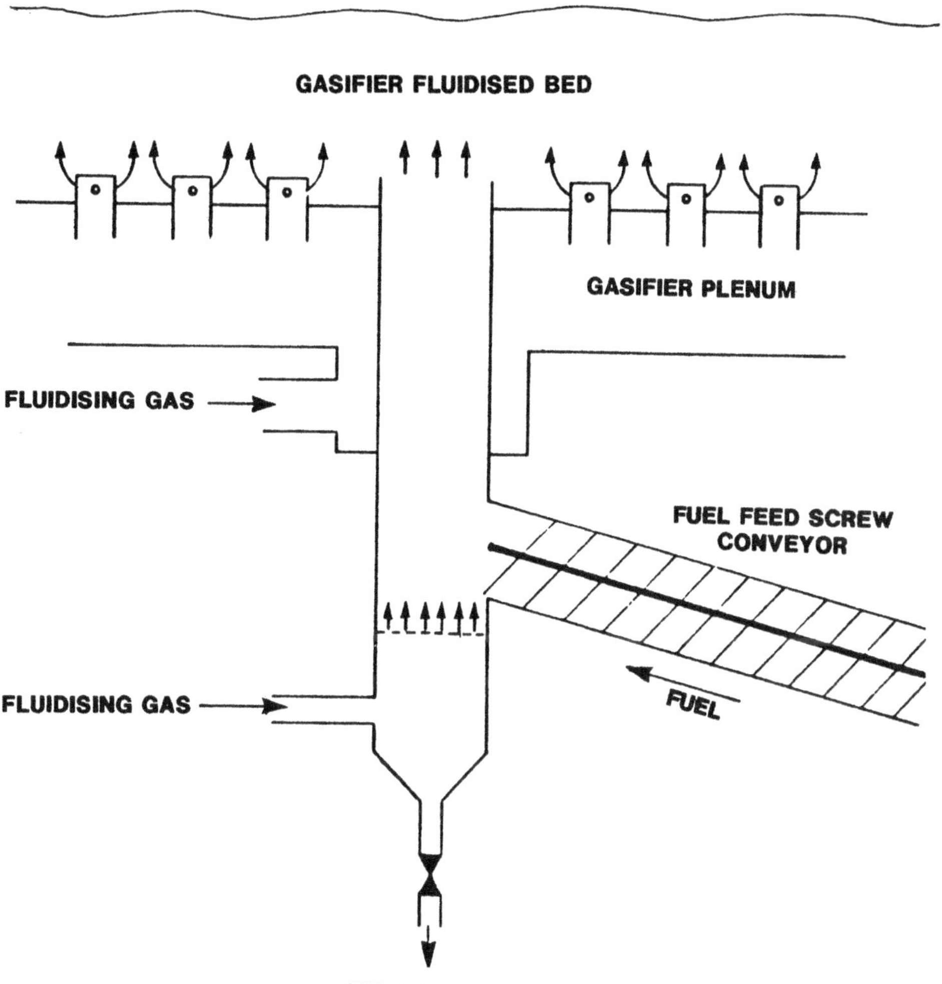

FIGURE 11

remove the majority of dust prior to quench cooling and scrubbing with water. The pressure of the cleaned gas was then boosted to 1350 mbar (abs) before splitting into product and recycle gas streams.

2.8.1 Gas Cyclone

The cyclone was of a high throughput design, having a mild steel casing lined with 0.15m of insulating castable and 0.1m of hard face castable refractory. The inlet nozzle was 3mm thick Incoloy 800H material.

The solids separated by the cyclone collected in a water jacketed dust catcher where they cooled prior to being discharged into open bins via a small lock hopper. Pneumatic actuated knife valves were used at the inlet and outlet of the lock hopper. Rotary paddle type level switches were used to indicate the level of solids in the dust catcher.

2.8.2 Gas Quenching and Scrubbing System

This comprised a primary and secondary wash column designed to cool the gas from 950°C to any desired temperature between 90°C and 20°C.

The hot gas leaving the cyclone was directed concentrically into the top of the primary wash column where it was immediately quench cooled by water sprays. Further water sprays cooled and scrubbed the gas in a section of open tower before liquid and gas travelled cocurrently through a section packed with Pall rings to bring the wash water and gas into thermal equilibrium. The clean gas left the base of the tower at approximately 40°C and entered the secondary wash column.

The sludge liquor stream passed from the wash column through an oil liquor separator and into a small surge tank fitted with make-up water supply and overflow drain. From the tank the wash water was pumped through a plate type heat exchanger where it cooled before returning to the wash column spray system.

The secondary wash column was designed to cool the gas to 20°C by direct contact with cold mains water flowing counter current to the gas over Pall rings. A demister element mounted at the column outlet removed entrained water droplets from the gas stream. The clean warm water collected at the base of the column was used in various locations around the plant.

By control of the water flow to each column the temperature of the gas leaving the system could be controlled at any value between 20°C and 90°C.

2.8.3 Gas Boosters

In order that the gas leaving the wash columns could be used to fluidise the gasifier it was necessary to boost its pressure from 1030 mbars (abs) to 1330 mbars (abs). This was achieved in two high speed centrifugal blowers connected in series. Labyrinth seals fitted with a nitrogen purge were used to prevent gas leakage between the shafts and fan casings.

All the gas leaving the wash columns passed through the boosters and the gas excess to that required for fluidising the gasifier (the product gas) was piped to an incinerator for disposal.

2.9 FLUE GAS CLEAN UP SYSTEM

The hot flue gas leaving the oxidiser passed through a cyclone, to remove the majority of dust prior to entering the tube side of a shell and tube type air preheater. The flue gas leaving the preheater at 550°C was discharged to atmosphere via a stack.

2.9.1 Flue Gas Cyclone

Both the cyclone and the solids discharge system were identical in size and construction to those used on the product gas stream and described in para 2.8.1.

2.9.2 The Air Preheater

This was a horizontally mounted shell and tube type heat exchanger containing 260, 25mm O.D. tubes and capable of preheating the fluidising air from 25°C to 500°C. Flue gas passed through the heat exchanger on the tube side and air passed counter current through the shell side. A valve in a bypass line joining shell inlet and outlet provided control of the air preheat temperature.

The complete unit was manufactured in type 310 stainless steel.

2.10 CONTROL AND INSTRUMENTATION

The plant was automated with all the controllers and indicators mounted in a control room local to the plant.

2.10.1 Main Control Loops

The main control loops are indicated on the oxidiser and gasifier E.L.D's, figures 12 and 13, the more important ones being:-

Pump Section and Fluidising Section Air

These were under individual flow control. An annubar

ODG - GASIFIER ENGINEERING LINE DIAGRAM

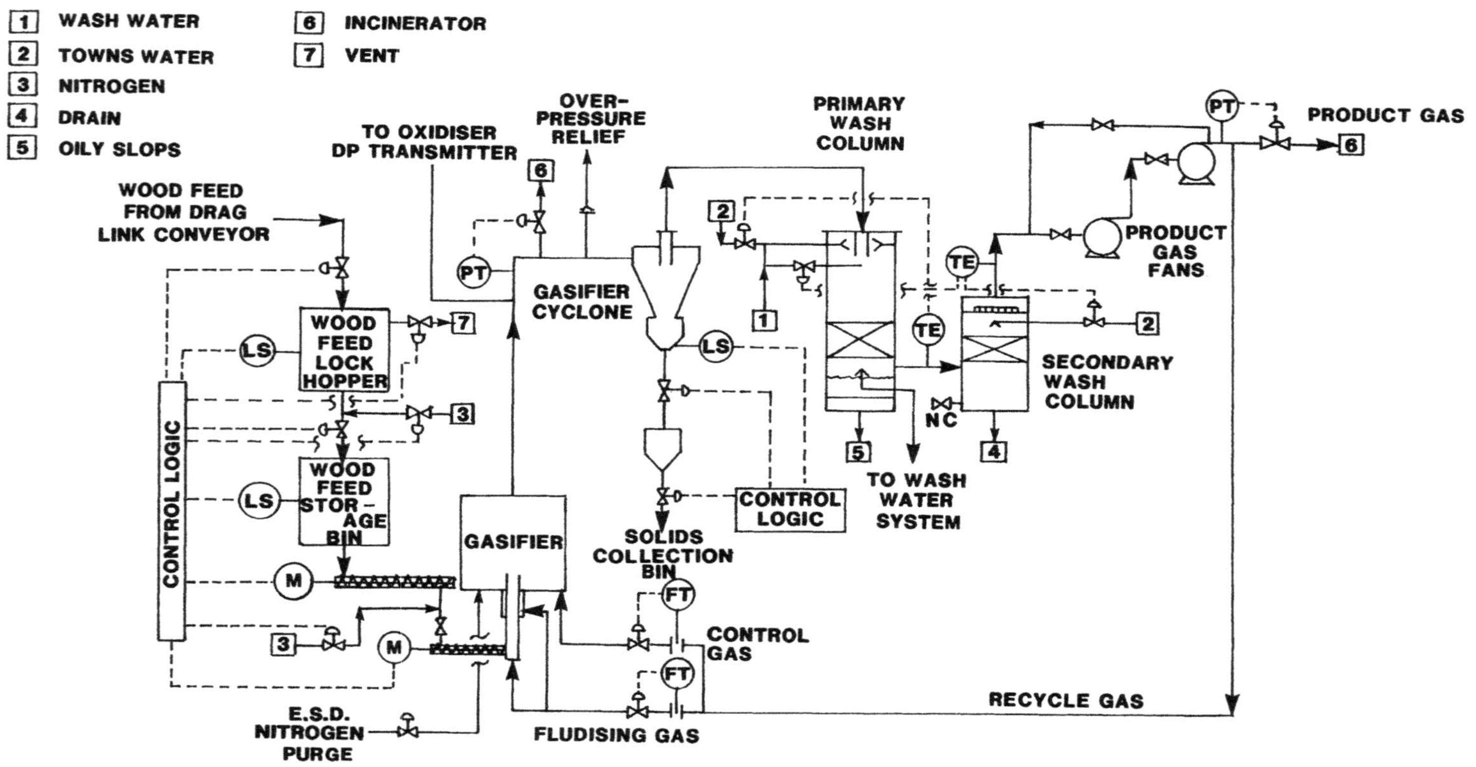

FIGURE 12

ODG – OXIDISER ENGINEERING LINE DIAGRAM

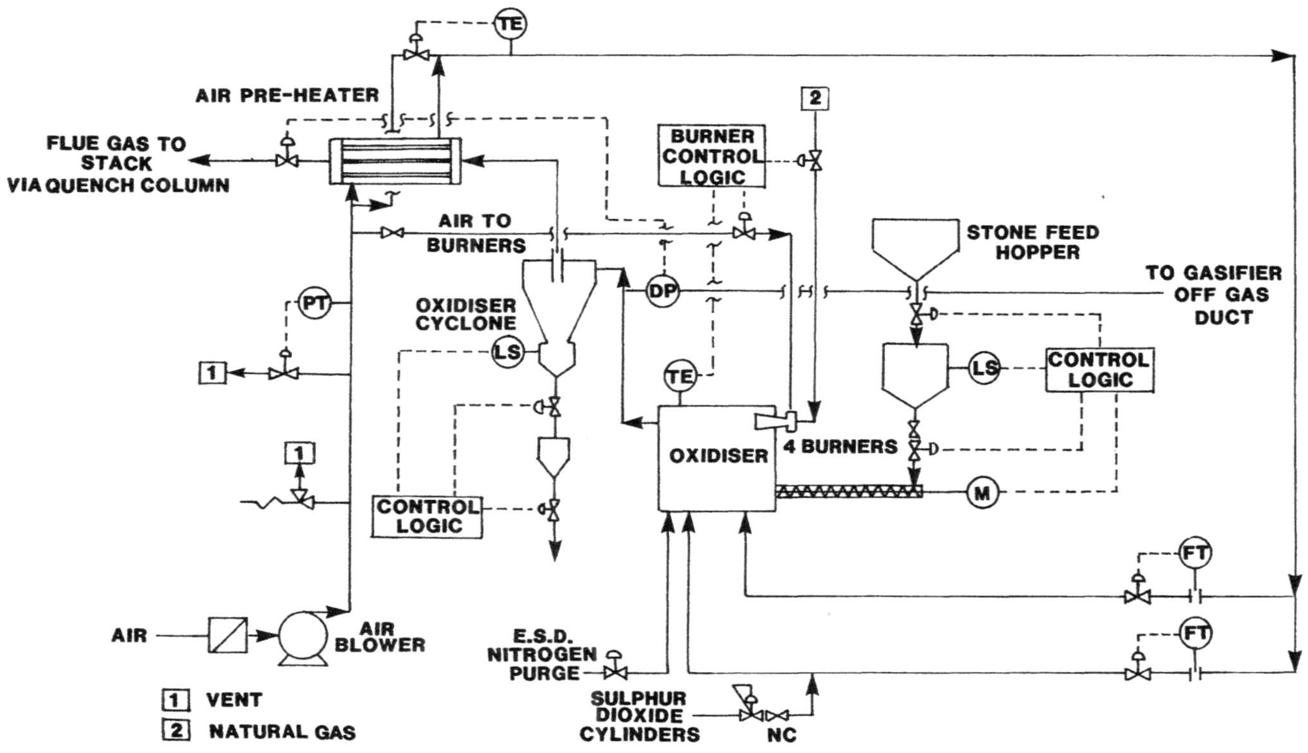

FIGURE 13

mounted in each line determined the air flow and downstream valves controlled each flow to the set point value.

Fluidising Air Pressure

A pressure controller maintained the air header at the set point pressure by venting excess air to atmosphere.

Fluidising Air Temperature

This was controlled by a valve on a bypass line between the air inlet and outlet of the air preheater.

Flue Gas Pressure

The differential pressure between oxidiser and gasifier outlets was maintained at a set point value by control of a valve in the flue gas line.

Pump Section and Fluidising Section Gas

The flow control system used was similar to that described for pump and fluidising air.

Fluidising Gas Pressure

When producing gas, this was maintained at a set point value by control of a valve mounted in the product gas line.

Product Gas Temperature

A split range instrument controlled the flow of water to both primary and secondary wash columns in order to cool the gas to the set point temperature.

Reactor Temperature

This was adjusted by varying the rate of fuel fed to the gasifier and oxidiser compartments.

Miscellaneous Control Systems

A P.L.C. was used to control the batching of solids from the gasifier and oxidiser dust catchers. Limit switches, actuated only when the batching valves were closed, ensured that the inlet and outlet valves could not be opened together.

Carbon Monoxide detectors were located where it was possible that faulty seals, on valves or fans, might lead to the accumulation of toxic/inflamable gases.

Audio-visual alarms warned the operator if temperatures, pressures, flows or levels moved outside predetermined bands. Upset conditions which reached a point of endangering the plant initiated an emergency shut down which automatically returned the plant to a safe condition by stopping the fuel feed and purging all the process equipment with nitrogen.

2.10.2 Data Collection

Annubars were used to measure product gas and flue gas flow rates. Product gas and flue gas compositions were determined by on line analysers and in addition samples of gas were periodically collected for analysis by gas chromatograph. The on line analysers were as follows :-

- Infra-red analysers measured carbon monoxide, carbon dioxide and methane concentration

- A thermal conductivity analyser measured hydrogen concentration

- Oxygen concentration was determined by measuring relative magnetic susceptibility

All the analysers were calibrated immediately before a test run and the gas analyses given by the on line analysers and by gas chromotography agreed within reasonable limits.

The temperature profile in the reactor was determined by sixteen thermocouples mounted in the oxidiser and nine in the gasifier.

During experimental runs, at ten minute intervals, a data logger recorded the information necessary for heat and mass balances to be determined. Log sheets which recorded the major operating parameters were completed at hourly intervals throughout the trial period.

3.0 OPERATION OF THE PILOT PLANT

On the 26 May 1984 the majority of construction work was complete and precommissioning had reached the stage where the natural gas burners could be lit and refractory dry out and curing started. The following 10 weeks were spent gaining operating experience and overcoming 'operational' problems. The first gas was produced on 10 August.

3.1 Operational Problems Encountered

These fell into 3 catagories:

- Minor recurring problems with proprietary items of equipment.

- Problems associated with feeding solids to a fluidised bed.

- Other problems.

3.1.1 Minor Recurring Problems

Valves Handling Solids

Throughout the plant knife blade slide valves were used

in pipework conveying solids. Generally in situations where the valves were not cutting through solids they worked satisfactorily but where they did, it was only a matter of time before the valve became unservicable due to compacted solids jamming the knife blade. This necessitated regular removal of the valves for cleaning. The problem was partly solved by mounting the valves so that any ledges in the vacinity of the seat faced downwards thus discouraging the build up of solids. In all cases, this resulted in the valve being mounted the opposite way round to that advised by the manufacturers but as the plant operated at a low pressure this did not present a problem.

Instrumentation

Rotary paddle type level switches mounted in the cyclone dust catchers were almost continually out of action due to very fine dust compacting around the shaft and eventually preventing its rotation. Numerous shielding devices were installed, but none were effective. Although never installed, a back purge of a clean gas around the rotating shaft and its bearings was thought to be the most promising solution to this problem.

Capacitance type level switches worked very well on wood chips but they could not be used in the dust catchers owing to temperature limitations.

The small pressure tappings in the annubar used to measure the flow of flue gas often became blocked with dust, these were easily cleaned by blowing them through with compressed air.

During gas production small quantities of naphthalene, crystallising in the gas analyser gas cooler pipework, had to be removed periodically.

Air Preheater

A quantity of the very fine dust, remaining in the flue gas after the flue gas cyclone, collected in the air preheater blocking the tubes and forming a honeycomb structure on the inlet tube plate. Little could be done to relieve this problem and the heat exchanger had to be cleaned weekly to prevent excessive pressure drop across the unit.

3.1.2 Problems associated with Feeding Solids to the Fluidised Bed

These presented the greatest problem and significant changes had to be made to the original design.

Stone Feeder System

The installed system satisfactorily fed material into

the empty oxidiser. However, once the fluidised bed reached the top of the screw discharge port the screw feeder tripped and it was impossible to feed further material. To overcome this problem the following modifications were made.

- A manual valve was installed directly above the screw feeder inlet so that a "trickle feed" of solids into the screw could be established allowing the screw to be operated at a high speed whilst almost empty. This proved reasonably successful but the rate at which material was fed to the screw was critical and the system was not reliable enough to maintain long periods of uninterrupted operation.

- The screw shaft and flights were extended in 310 stainless steel so that the screw reached the inside wall of the oxidiser refractory and hence the material being fed did not have to be extruded the last 150mm into the fluidised bed. This modification had little effect.

- The variable speed 1.5 KW drive unit was replaced by a 4 KW fixed speed unit. Initially this proved successful but after several days of continuous difficulties, with the screw becoming jammed and even stripping its flights the complete feeder was raised 0.5m so that material was being fed onto the surface of the oxidiser bed. Later the 1.5 KW drive unit was replaced and apart from minor problems the system worked satisfactorily for the remainder of the trials.

Fuel Feed System

After the experience gained with the stone feeder it was not unexpected when the fuel feeder, which was also operated by screw feeding material in to the fluidised bed, failed. There were two reasons why it was judged that the problem would not be overcome by feeding fuel above the bed surface, these were:-

- The wood chip being of lower density than the bed material would tend to float on the surface of the bed and not become mixed with it. Pyrolysis of the fuel on the surface of the bed in the absence of oxygen was likely to result in the gas containing unwanted tars.

- Owing to the position of the gasifier within the oxidiser, it was very difficult to feed material into the gasifier, except upwards through the plenum chamber.

It was anticipated, however, that by substantially

increasing the gas velocity in the feed tube, fuel would be entrained into the fluidised bed above. To this end, the sieve type distributor plate in the base of the feed tube was replaced by nozzles. In practice, limited by entrainment of material from the gasifier, it was not possible to carry the wood chip into the fluidised bed; instead it collected in the feed tube quickly forming a blockage.

The fuel feed system was eventually removed from the reactor and a test rig built to investigate better methods of feeding fuel into the base of a fluidised bed.

The system developed was as shown in figure 14. The feed tube was divided equally by a central baffle plate and the segment on the fuel input side extended by 0.6m so that its exit was above the fluidised bed. A split plenum with nozzle type distributors was used to independently fluidise material on both sides of the baffle. By control of the fluidising gas flows it was possible to establish a rapidly pulsating flow of bed material down the short section of the tube and up the longer section to be deposited on the surface of the bed. Fuel fed to the base of the feed side of the tube was entrained into the circulating bed material. The system fed wood chip on to the surface of the bed mixed with hot bed material.

Operation of this system at temperature in the reactor presented two problems:-

- There were difficulties in obtaining reliable gasket material for the flanged joints at the base of the feed tube assembly which were operating at 950°C.

- On feeding fuel to the system the gas velocity in the feed side of the tube increased dramatically and material was ejected into the gasifier offtake duct where the velocity was sufficient to cause bed material and wood char to become entrained.

Under these conditions short periods of operation were possible and some poor quality gas was produced. Attempts at controlling the system were unsuccessful owing to the very rapid rate at which material was removed from the reactor once entrainment occured. Eventually a deflector plate was fitted to the top of the tube so that the material leaving was deflected horizontally. This worked well and the feed tube could be operated for long periods with negligible carryover of bed material from the gasifier into the downstream equipment.

Having established an operational feed tube system the next series of problems were associated with the screw conveyors. Wood chip tended to bridge in the outlet

DETAILS OF Mk.2 FUEL FEED TUBE

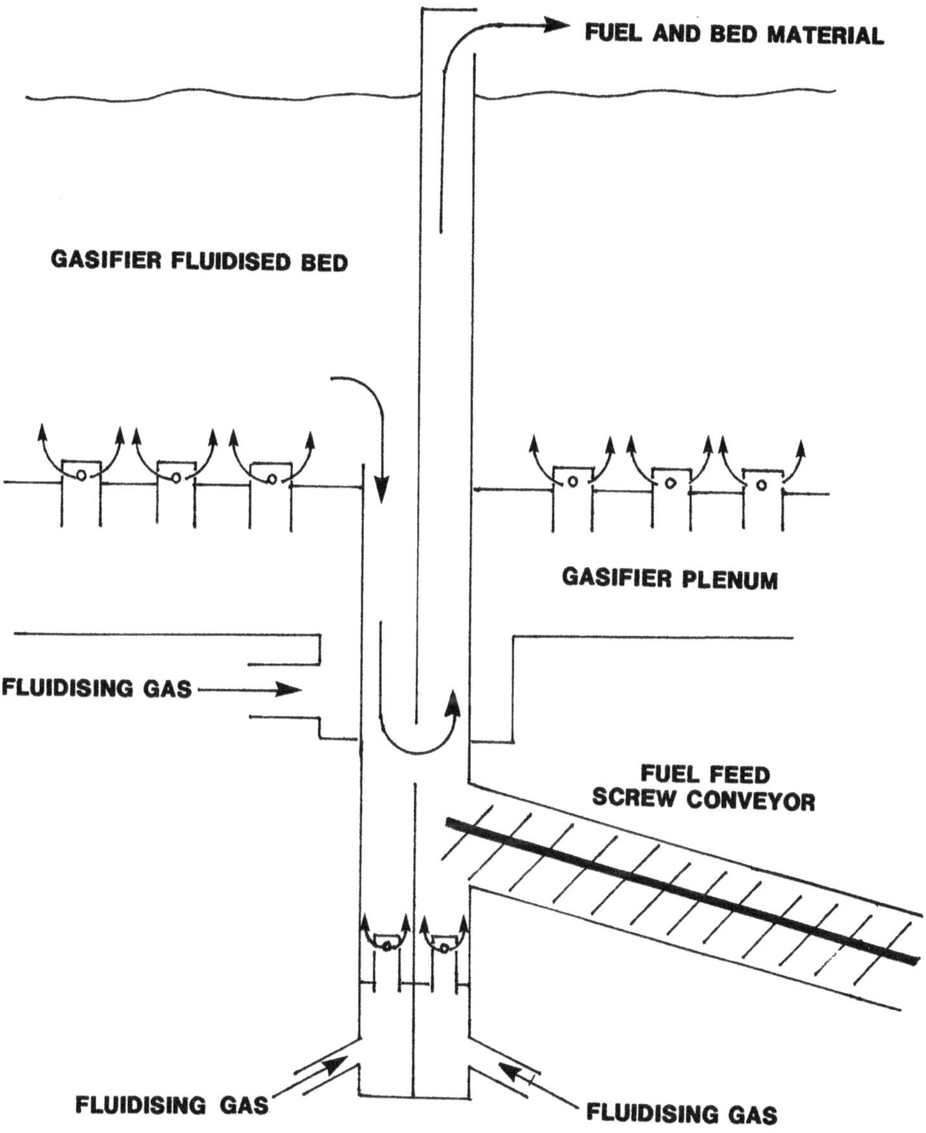

FIGURE 14

pipe of the first conveyor blocking its exit. This resulted in the wood chip being fed to the unit becoming compacted in the end of the screw and jamming it. In an attempt to overcome this problem the following modifications were made:-

- A pressure balancing line was installed across the screw conveyors.
- A small but positive flow of nitrogen was established along the screws and into the gasifier.
- The isolation valve between the screw conveyors was removed.
- The flights at the exit of the first screw conveyor were modified to discourage material from being compacted at the end of the screw.

These changes produced little improvement and subsequent plant shut downs generally resulted from the above feed problem.

3.1.3 Other Problems

Owing to the low relief pressure of the oxidiser explosion panels they were frequently ruptured by inadvertantly over pressurising the system. The cause was never serious but replacing the panels and bringing the plant back to operating temperature took several hours. The explosion panels were never required to function for their intended purpose of relieving the oxidiser during an explosion.

Initially during start up an attempt was made to use wood chip to heat the system from 500°C to 950°C. Unfortunately the wood tended to be carried out of the fluidised bed and burn in the free board thereby only heating the flue gas. Various techniques to keep the wood in the bed were tried but the highest bed temperature attained was 750°C. This problem was overcome by feeding coal to the oxidiser which burnt within the bed causing its temperature to rise to the operating condition in a controlled manner.

On one occassion during start-up high temperatures in the oxidiser resulted in melting a section of the gasifier shell, five distributor nozzles and two thermocouple sheaths. The reason for this high temperature excursion was considered to be due to feeding coal on to the surface of a slumped bed which burnt vigorously in a localised area when only the oxidiser bed was fluidised. The damage was repaired and an operating procedure established to prevent a recurrence of the problem.

3.2 OPERATING PROCEDURE

3.2.1 Start-Up

With a sufficient depth of bed material in the reactor the oxidiser was gently fluidised with air for several minutes to purge the oxidiser and flue gas clean up equipment. The start up burners were then lit and the flue gas temperature increased to 900°C at the required rate. As the flue gas temperature rose the air preheat temperature gradually increased and in turn so did the bed temperature. When the oxidiser bed reached 450°C the gasifier was gently fluidised using air pressurised by the gas boosters. With the gasifier and oxidiser beds fluidised the bed material circulated and attained a uniform temperature. When all the bed material reached 500°C coal was slowly fed to the oxidiser through the stone feed unit and the bed temperature rose rapidly. The start up burners were extinguished and the coal feed rate was adjusted so as to increase the reactor to its operating temperature of 950°C at the desired rate.

Once the start up procedure had been established the reactor could be brought up to operating temperature from cold in 12 hours.

3.2.2 Bed Material Conditioning

To produce the required bed material the limestone had to be calcined and partially sulphated during the initial start-up period. This was achieved by maintaining the bed at a temperature between 800 and 850°C and fluidising the oxidiser with air which contained sulphur dioxide.

3.2.3 Normal Operation

Having brought the reactor and a sufficient quantity of active bed material upto the operating temperature of 950°C a flow of bed material through the redesigned fuel feed tube was established. The air inlet to the gas boosters was isolated, thus recycling the gas in the system to fluidise the gasifier. Woodchip was then fed at the desired rate and the pressure in the gasifier was controlled by manual adjustment of the valve in the product gas line. Within minutes of feeding fuel to the gasifier the gas composition changed from high levels of carbon dioxide to moderate levels of carbon monoxide.

The temperature of the oxidiser was maintained at about 1000°C by controlling the rate at which coal was fed to the oxidiser. For a constant fuel feed rate both the gas composition and production rate quickly stabilised.

3.2.4 Shut Down

The fuel feed to the gasifier was stopped and the

gasifier pressure maintained by closing the product gas valve. As the concentration of carbon in the gasifier fell the combustable components of the gas were oxidised to water and carbon dioxide which resulted in the formation on an inert recycle gas. Once the recycle gas was inert, air was allowed to enter the gas boosters and fluidise the gasifier. With no fuel being fed to the reactor its temperature quickly dropped and on reaching 500°C the fluidising air and gas were stopped and the bed allowed to slump.

3.3 WOOD GASIFICATION TRIALS

Sixteen separate wood gasification runs were performed but because of the problems with the fuel feed system those initially carried out lasted only a few minutes and produced little useful data. The final four runs carried out using the fuel feed tube with its outlet deflector lasted a reasonable time and provided consistent data. The longest run lasted 9 hours 43 minutes and the data collected enabled a mass and energy balance to be established. Failure of the flue gas flow meter during this run prevented a true experimental mass balance to be made. Flue gas flow rate had to be calculated by mass balance over the reactor. Data collected in runs 14, 15 and 16 indicated that a good quality gas was generated soon after feeding fuel to the gasifier and also confirmed the mass and energy balance derived for run 13.

3.3.1 Experimental Results

The most useful experimental data collected over runs 13 to 16 is given in tables 2 to 5.

3.3.2 Analytical Results

All analytical work was performed by BCRA Scientific and Technical Services Limited based in Chesterfield, England.

Wood Analysis

The wood used had been chipped and partially dried. It had a typical analysis after drying of :-

Proximate Analysis (as received basis)

Moisture	19.1 wt%
Ash	0.6 wt%
Volatile Matter	68.0 wt%
Fixed Carbon	12.3 wt%

Ultimate Analysis (dry, ash free basis)

Carbon	50.8 wt%
Hydrogen	5.9 wt%
Nitrogen	0 wt%
Oxygen	43.27 wt%
Sulphur	0.03 wt%

TABLE 2

RUN NO. 13 START TIME 23:30 10/8/84 DURATION 9 HOURS 34 MINTUES

TIME	WOOD FEED RATE kg/h	COAL FEED RATE kg/h	GAS COMPOSITION VOL %				PRODUCT GAS RATE kg/h	RECYCLE GAS RATE kg/h	OXIDISER AIR RATE kg/h	BED TEMP °C	
			CO	CO_2	H_2	CH_4				ox-in	ox-out
00:00		0	29.5	25.0	20.3	9.5			1412	884	935
00:21		0	33.0	20.0	26.4	11.0					
01:00	105	0	31.9	16.1	24.6	12.4	199	428	1386	852	892
01:46		0	30.8	15.2	24.6	13.0					
02:00		0	32.5	15.5	23.8	12.0	200	429	1530		
02:30	447	0	33.7	14.4	26.7	12.4					
03:00	338	43.6	35.0	14.5	25.8	12.6	212	423	1530	867	909
03:25			36.0	13.0	28.8	11.0					
04:00	240	43.6	36.5	11.8	28.3	11.3	207	417	1530		
04:38	569		36.4	9.9	28.0	11.6					
05:00	626	43.6	36.2	9.8	27.8	12.1	262	414	1530	878	922
05:31	522		37.8	14.4	32.5	12.2					
06:00	250	43.6	37.0	14.1	34.6	12.6	208	405	1530		
06:34	321		37.8	13.9	33.6	11.6					
07:00		0	37.7	12.5	31.5	12.4	195	409	1530	904	955
07:37	275		37.6	15.6	32.1	12.7					
08:03		43.6	36.1	17.0	30.4	12.3	208	417	1530		
08:48	417		37.8	15.9	31.9	12.7					
09:06		43.6	37.3	16.0	31.3	12.7	197	414	1530	882	917
09:24			37.5	17.9	29.2	12.3					

Gasifier pressure maintained at 30 mbar above oxidiser pressure throughout run.
Bed depth = 420mm

TABLE 3

RUN NO. 14 START TIME 7:48 16/8/84 DURATION 16 MINUTES

TIME	WOOD FEED RATE kg/h	COAL FEED RATE kg/h	GAS COMPOSITION VOL %				PRODUCT GAS RATE kg/h	RECYCLE GAS RATE kg/h	OXIDISER AIR RATE kg/h	BED TEMP °C	
			CO	CO_2	H_2	CH_4				ox-in	ox-out
7:52		0	4	>30	-	1.6					
7:55		0	24	25	-	5.6					
7:57		0	32	18	-	8.8					
8:00	240	0	34	15	25.8	>10.0	127				
8:03		0	38	17	26.0	9.0					

TABLE 4

RUN NO. 15 START TIME 5:16 17/8/84 DURATION 2 HOURS 15 MINUTES

TIME	WOOD FEED RATE kg/h	COAL FEED RATE kg/h	GAS COMPOSITION VOL % CO	CO_2	H_2	CH_4	PRODUCT GAS RATE kg/h	RECYCLE GAS RATE kg/h	OXIDISER AIR RATE kg/h	BED TEMP °C ox-in	ox-out
5:16		0									
5:25			38	21	23.8	5.2					
5:35			34	19	25.6	9.5	155	459	1386	845	933
5:50			32	18	24.8	10.0					
6:02		0	30.5	18	24.8	>10.0	163	460	1386		
6:16	342									776	834
6:36			30	16	26.7	>10.0					
6:45	0										
6:50	80										
6:57	242										
7:05		0	22	13	22.9	>10.0	70	457	1386	758	826
7:22	342										
7:35			33.4	14.3	28.1	>10.0					

Bed Depth = 320mm

TABLE 5

RUN NO. 16 START TIME 21:16 17/8/84 DURATION 1 HOUR 21 MINUTES

TIME	WOOD FEED RATE kg/h	COAL FEED RATE kg/h	GAS COMPOSITION VOL % CO	CO_2	H_2	CH_4	PRODUCT GAS RATE kg/h	RECYCLE GAS RATE kg/h	OXIDISER AIR RATE kg/h	BED TEMP °C ox-in	ox-out
21:18	165	0								837	855
21:23			14.0	23.0		4.5					
21:25	240		19.5	26.0		6.8					
21:30			29.0	24.0	23.8	9.4	151	498	1419		
21:42			30.5	23.0	25.8	9.5				848	867
21:52			30.0	22.5	24.7	>10.0					
22:00	240						155	537	1419		
22:11		0	28.8	24.0	23.7	>10.0					

Gross Calorific Value = 16,160 kJ/kg (as received)

Sample Collected 10.9.84.

Coal Analysis

The coal used to heat the oxidiser was smalls from Littleton Colliery. Only material smaller than 1 cm^3 was fed to the oxidiser.

Proximate Analysis (as received basis)

Moisture	17.3 wt%
Ash	10.7 wt%
Volatile Matter	28.2 wt%
Fixed Carbon	43.8 wt%

Ultimate Analysis (dry ash free basis)

Carbon	81.51 wt%
Hydrogen	5.4 wt%
Nitrogen	1.71 wt%
Oxygen	10.13 wt%
Sulphur	1.25 wt%

Gross Calorific Value = 24,280 kJ/kg (as received)

Limestone

The limestone forming the basis of the bed material in the successful trials was a high purity sedimentary limestone quarried from Middleton Quarry, Derbyshire and had an analysis before calcinations :-

Moisture	1.0 wt%
Carbon Dioxide	43.7 wt%
Calcium Oxide	52.1 wt%
Silica	0.4 wt%
Iron Oxide (Fe_2O_3)	0.02 wt%
Magnesium Oxide	0.26 wt%

Size Distribution :-

0.76 mm - 0.57 mm	70.0 wt%
<0.57 mm	30.0 wt%

Bed Material

The composition of the bed material at the start and end of the trial period which included runs 13 to 16 was :-

	29/7/84	21/8/84
Calcium Oxide	66.0 wt%	92.6 wt%*
Calcium Carbonate	20.0 wt%	5.9 wt%
Calcium Sulphate	8.9 wt%	1.5 wt%
Ash	5.1 wt%	-

* includes ash

Cyclone Fines

The composition of the fines collected from the gasifier and oxidiser cyclones is given in Table 6.

Typical size distribution of the fines was :-

>800 µm	0.7 wt%
600 - 800 µm	13.5 wt%
422 - 600 µm	33.5 wt%
250 - 422 µm	36.3 wt%
<250 µm	16.0 wt%

3.3.3 General Comments

During the initial heating and calcination of the limestone considerably more fines were generated than when the bed was fluidised under similar conditions in its sulphated state. Over periods of stable operating the quantity of fines generated were slightly less than had been expected and the cost of make up stone required by the system is not considered a problem.

Gas quality and production were dependent on a steady supply of fuel to the gasifier. An interruption in the fuel feed resulted in an immediate loss of gasifier pressure and a reduction in the quantity of carbon monoxide and hydrogen in the product gas. During gas production the gasifier pressure was maintained between 30 and 50 mb above that in the oxidiser and during periods of steady fuel supply the variation in differential pressure was usually less than 5 mbar. For fuel feed rates below 200 kg/h the concentration of carbon monoxide and hydrogen in the product gas could not be maintained at the levels produced at higher fuel feed rates.

Owing to the difficulties in supplying woodchip to the gasifier fuel feed tube, an average feed rate in excess of 340 kg/h was never achieved.

Irrespective of the fuel feed rate only very low levels of carbon monoxide and hydrogen were produced when the bed temperature was less than 850°C but the levels increased very rapidly once the temperature reached 850 - 860°C.

There was no evidence of oils or tars in the gas cyclone or wash columns. However, at the end of the trials the wash water had the odour of naphthalene and the base of the Primary Wash Column contained a quantity of very fine sludge consisting of calcium sulphate/oxide. It is not known whether the naphthalene originated from the wood fed to the gasifier or from the coal used to initially heat up the reactor.

TABLE 6

CYCLONE FINES ANALYSIS

CYCLONE	DATE	RUN NO.	CALCIUM SULPHATE	CALCIUM SULPHIDE	CALCIUM CARBONATE	CALCIUM OXIDE	ASH	CARBON
OXIDISER	29/7/84		5.5	-	20.0	62.7	7.9	3.9
GASIFIER	29/7/84		6.2	2.9	32.7	47.4	6.8	4.0
GASIFIER	11/8/84	13	1.2	0.8	6.0	73.3		5.7
OXIDISER	17/8/84	15	3.7	-	7.2	71.5		2.6
GASIFIER	17/8/84	15	1.3	0.6	16.5	55.5		11.1

All Figures wt%

3.4 MECHANICAL PERFORMANCE OF REACTOR

The Incoloy gasification chamber was subjected to a particularly arduous environment having to operate at temperatures in excess of 1000°C in both a reducing and oxidising atmosphere. In addition, owing to the nature of experimental work, the chamber was subjected to frequent heating and cooling cycles which resulted in upto a 50 mm change in the vessel height. At the end of the trials the surface of the Incoloy was slightly pitted and had a thin coating of brown scale. The upper section of the chamber remained in a good condition but the section in the fluidised bed was buckled though generally sound.

Although in service for only a few weeks the fuel feed tube assembly manufactured in type 310 stainless steel showed no signs of distortion or wear. Again a thin brown coating had formed on the metal surface.

Apart from a single short crack in a butt weld the plenum chamber and distributor nozzles had no visible signs of wear at the end of the trials.

As expected with castable refractory, over the dryout and curing period, the oxidiser lining developed several thin cracks but these were not a problem.

4.0 ASSESSMENT OF RESULTS

The data collected enabled a representative gas composition to be determined together with overall mass and energy balances. Unfortunately, owing to time restraints, insufficient data was collected to produce a sulphur balance over the reactor.

4.1 PRODUCT GAS COMPOSITION

A plot of variation in product gas composition throughout run 13 is given in Figure 15. The gas composition at steady state operation was taken as being :-

Carbon Monoxide	37.5 vol%
Carbon Dioxide	14.5 vol%
Hydrogen	31.0 vol%
Methane	12.2 vol%
Nitrogen	4.8 vol%

Gross Calorific Value = 13.58 MJ/Nm^3 @ 0°C.

Since the gasifier was always maintained at a higher pressure than the oxidiser, the nitrogen present in the gas most likely originated from the nitrogen purges used to prevent gas leakage from the gas booster shafts and from the nitrogen introduced into the fuel feed system to combat bridging.

VARIATION IN PRODUCT GAS COMPOSITION DURING RUN 13

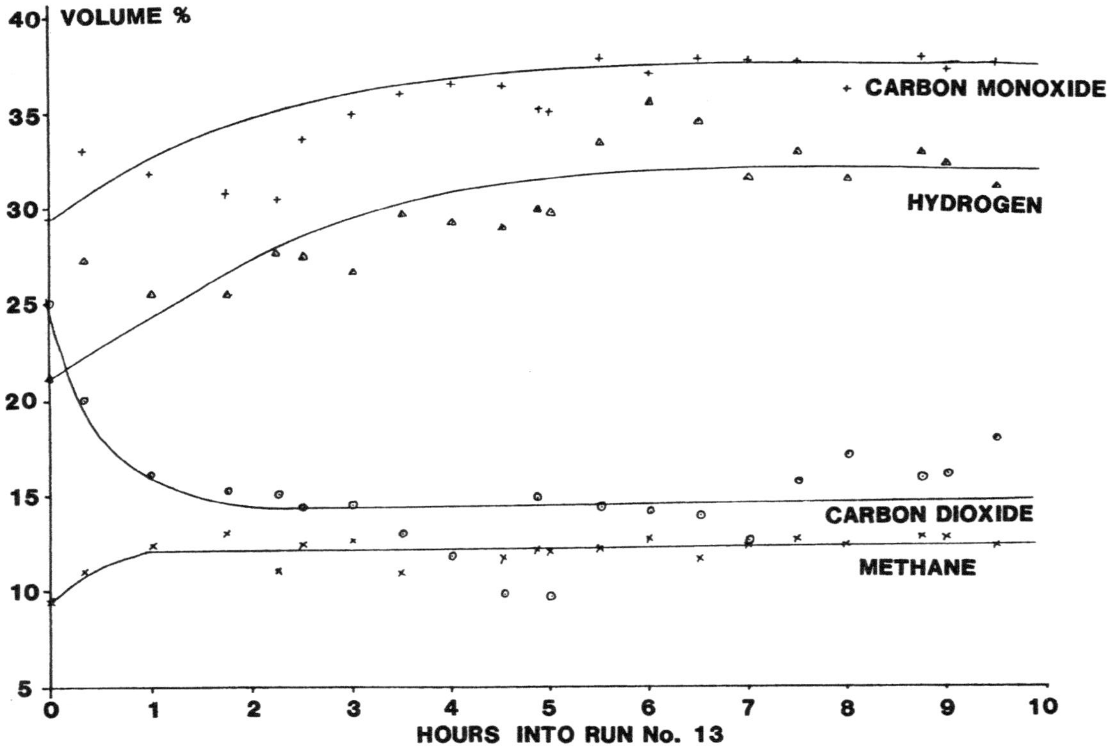

FIGURE 15

4.2 MASS BALANCE

The overall mass balance is given in Table 7. The only component in which the quantity or composition were not actually measured was the flue gas and this was deduced assuming that all gaseous elements entering the system left the system. The flue gas thus formed the balance between gaseous elemental inputs and outputs and had a theoretical composition :-

Carbon Dioxide	13.0 vol%
Water Vapour	13.7 vol%
Nitrogen	69.0 vol%
Oxygen	4.3 vol%

As this was similar to on line analytical spot checks on flue gas compositions, the mass balance was assumed to be satisfactory.

Several assumptions had to be made in order to produce the mass balance, these were :-

- The water content of the gas leaving the gasifier was negligable.
- The gas leaving the secondary wash column was saturated at 30°C.
- The air used to fluidise the oxidiser was completely dry.
- Tars, Oils and Liquors were assumed to be negligible.
- No allowance was made for solids not collected by the cyclones.
- The system was operating under steady state conditions and carbon was not accumulating in the gasifier.

The mass balance indicates that ash builds up in the bed at a rate of 4.65 kg/hour which is considerably faster than that deduced from the ash content of the bed at the end of the trials.

The mass balance gives the oxygen to wood ratio as about 0.9:1 and not the theoretically predicted value of 0.6:1. This occurs because air was supplied to the oxidiser through the start up burners at a rate of 450 Kg/h in order to prevent the accumulation of bed material in the burner nozzles. When this occurs this excess air is taken account of, the oxygen/wood ratio reduces to the expected 0.7:1.

TABLE 7

MASS BALANCE (kg/h)

COMPONENT	CARBON	HYDROGEN	NITROGEN	OXYGEN	ASH	CALCIUM OXIDE	TOTAL
WOOD	139.45	23.45		176.81	2.21		342.00
AIR			1,175.04	354.96			1,530.00
WATER		2.51		20.09			22.60
COAL	26.74	2.47	0.56	8.96	4.86		44.00
LIMESTONE	6.6			17.60		30.80	55.00
TOTAL INPUT	172.79	28.43	1,175.60	578.42	7.07	30.80	1,993.11
PRODUCT GAS	76.92	11.94	13.37	113.27			215.50
GASIFIER FINES	1.32			0.42	1.16	16.49	19.39
OXIDISER FINES	0.91			0.65	1.78	18.96	22.30
FLUE GAS	93.64	16.49	1,162.23	464.08			1,736.44
TOTAL OUTPUT	172.79	28.43	1,175.60	578.42	2.94	35.45	1,993.63

4.3 ENERGY BALANCE

Based on the experimental data the energy balance over the reactor was :-

INPUT		GJ/h	OUTPUT		GJ/h
Wood	CV	5.528	Gas from Cyclone	CV	8.908
Preheated Air	SH	0.433		SH	0.884
Recycle Gas	CV	5.869			
	SH	0.035	Flue gas to		
			Preheater	SH	1.746
Coal	CV	1.068		LH	0.335
			Gasifier Cyclone		
			Fines	CV	0.043
				SH	0.016
			Oxidiser Cyclone		
			Fines	CV	0.030
				SH	0.017
			Heat Losses		0.954
TOTAL		12.933	TOTAL		12.933

CV = Calorific Value (Chemical Energy)
SH = Sensible Heat
LH = Latent Heat

Basis :- 25°C, water as liquid.

The Energy balance was based on the mass balance presented in Table 7 together with the following measured data :-

- Preheated air entered the oxidiser at 420°C
- Recycled gas entered the gasifier at 100°C
- Flue gas and fines left the oxidiser at 875°C
- Gas and fines left the gasifier at 830°C

The total energy output was made to balance with the total energy input by the addition of heat losses. These amounted to 0.954 GJ/h representing 13.5% of the energy entering the reactor as sensible heat together with the chemical heat in the wood and coal. From measurements of the reactor outside wall temperature the expected heat losses would be 0.25 GJ/h which implies an error of 0.7 GJ/h in the energy balance.

4.4 THERMAL CONVERSION EFFICIENCY

The thermal efficiency of a process is dependent on how the energy generated in the process can be used. The thermal efficiency has thus been calculated for the following three cases:-

- Only the product gas chemical energy is of use.
- The product gas chemical energy together with the sensible heat in the product gas and flue gas is of use.

- The product gas chemical energy together with 75% of its sensible heat and 50% of the flue gas sensible heat is of use. This is the system considered most representative of an installation using an O.D.G.

The thermal efficiencies calculated in the above manner are:-

Energy Output	Energy Input	
	Wood C.V.	Wood and Coal C.V
Product gas C.V	55.0%	46.1%
Product gas C.V + Product gas S.H + Flue gas S.H	94.7%	79.4%
Product gas C.V + 75% Product gas S.H + 50% Flue gas S.H	78.8%	66.1%

4.5 POSSIBLE REACTION MECHANISM

The carbon, hydrogen and oxygen balance over the gasifier presented in Table 8 indicates that there is a net transfer of carbon, hydrogen and oxygen from the gasifier to the oxidiser.

TABLE 8 Mass balance over gasifier (kg/h)

	Carbon	Hydrogen	Oxygen
Input			
Wood	139.45	23.45	176.81
Water Vapour in Recycle Gas		2.51	20.09
TOTAL	139.45	25.96	196.90
Output			
Product Gas	76.92	11.94	113.27
Gasifier Dust	1.32		0.42
TOTAL	78.24	11.94	113.69
Input - Output	61.21	14.02	83.21

Part of the carbon transfer can be explained as wood char but owing to wood size, reactor temperature and residence time it is unlikely that fuel transferred to the oxidiser contains carbon, hydrogen and oxygen in the form of volatile matter. It is therefore a reasonable assumption that the hydrogen and oxygen together with some of the carbon must transfer to the oxidiser in a gaseous form. This hypothesis is supported by the fact that during gas production the gasifier was always maintained at a higher pressure than the oxidiser and this would encourage transfer of gas through the transfer slots into the oxidiser.

There are three possible explanations for the apparent discrepancy in the mass balance over the gasifier, these are :-

(i) Assuming that the gas transferred from the gasifier to the oxidiser was similar in composition to the product gas the transfer of carbon, hydrogen and oxygen can be broken down into 158.4 kg of product gas and 4.7 kg of carbon which leaves 5.2 kg of hydrogen unaccounted for. If this quantity of hydrogen is considered to be within experimental error then the transfer of oxygen from the oxidiser to the gasifier in the form calcium sulphate is unnecessary. In this case the overall system can be considered as a water gas reactor where the temperature in the gasifier is maintained by the heat in the circulating bed material.

(ii) It is possible that the hydrogen is present as water vapour in the product gas and the assumption that the product gas was dry is incorrect. If this is the case then 47 kg/h of oxygen is transferred from the oxidiser to the gasifier as calcium sulphate by the O.D.G. process to achieve a mass balance.

(iii) It could be assumed that the gas passing through the transfer slots had already been partially oxidised in the gasifier. This could be explained by considering the gasifier fluidised bed as having an oxidising lower section and a reducing upper section, the argument being supported by material densities. The least dense wood char, fed to the top of the bed would remain near the surface while the dense calcium sulphate rich particles would have a tendancy to migrate to the lower section of the bed. Recycled gas entering the gasifier would be partially oxidised to carbon dioxide, and water vapour some of which would be driven through the transfer slot into the oxidiser while the remainder entered the reducing section of the gasifier bed.

Assuming that 15 kg/h of carbon is transferred from the gasifier to the oxidiser as char and that the gas transferred to maintain the mass balance has been completely oxidised to carbon dioxide and water, the rate of oxygen transfer to the gasifier as calcium sulphate equates to 166 kg/h.

The first hypothesis assumes 5.2 Kg of hydrogen (ie, 20% of the hydrogen entering the system) is experimental error. This is extremely unlikely as the analytical results were consistent.

Both the first and second hypotheses depend upon the passage of medium calorific value gas through the transfer slots. This gas, burning on contact with air in the oxidiser, would be expected to generate intense heat, however, during gas production, the temperature in the region of both transfer slots was never noticably higher than the general bed temperature.

The final hypothesis appears to give the most probable explanation. Evidence that oxygen was transferred from the oxidiser to the gasifier as calcium sulphate is provided by the presence of calcium sulphide in the gasifier cyclone fines. Unfortunately, lack of time prevented the collection and analysis of bed material for calcium sulphide during gas production trials.

Also of interest is the bed temperature at which gas was produced. Until the gasifier bed attained a temperature in excess of 850°C very little gasification occurred. If the "water gas" reactions were alone responsible for gasification it would have been expected to have produced a reasonable quality gas at temperatures below 850°C.

4.6 ENVIRONMENTAL IMPACT

At this stage in the development of the ODG Process, whilst the nature of the wastes which may arise can be defined, their extent cannot and, neither therefore can the means necessary to treat and dispose of them.

Gaseous Wastes

The flue gas from the oxidiser is hot and depending on how the system is operated may contain sulphur dioxide as an undesirable component. It is likely that heat will be extracted from the flue gas and a relatively cool gas will be discharged from the process. There is the possibility of designing the oxidiser so that sulphur leaves the system in a sulphur dioxide rich side steam of flue gas. The sulphur dioxide concentration in this side stream is likely to be in excess of five volume percent which would make the recovery of sulphur economically viable. Further possibilities include the recovery of elemental sulphur from the system by reducing the sulphur dioxide to sulphur using a layer of carbon floating on the surface of the oxidiser fluidised bed. This provides a relatively simple and cheap method of removing and recovering sulphur from high sulphur containing feedstocks and is clearly an attractive concept worthy of further study.

Aqueous/Liquid Wastes

Organis wastes in the form of tars and oils are unlikely to occur in the process and aqueous wastes will only arise if the product gas is water cooled directly. An aqueous waste is likely to contain traces of organics with the main constituents being phenol and ammonium compounds. This should be a relatively easy material to purify by conventional means.

Solid Wastes

These may comprise heavy clinker, light ash and degraded fluidised bed materials. The clinker will be inert and because of its density, relatively easy to handle and dispose of. The light ash and dust collected from the cyclones may contain limestone, free calcium oxide, calcium sulphide, calcium sulphate, carbon and inert mineral matter. The oxide and sulphide will have to be treated before disposal and treatment with water appears an obvious option.

However, the sulphide would generate hydrogen sulphide and such treatment would therefore need to be carried out in a controlled manner.

All these matters will require attention in the latter phases of project development.

5.0 CONCLUSIONS

The overall aim of producing a medium calorific value gas from wood without the use of liquid oxygen was achieved. The mechanism by which gasification occurs is still uncertain but it is likely that the donation of oxygen from calcium sulphate plays a role in the gasification process.

The system produced a good quality tar free fuel gas but due to the high methane content it would be unsuitable for methanol production.

The maximum fuel feed rate achieved was 626 kg/h and the average fuel feed rate over the longest test run was 342 kg/h. This fuel feed rate was limited by problems with the wood feed systems and was not the rate of reaction in the gasifier. Similarly, the termination of test runs was due to the failure of the wood feed system, wood bridging and compacting being the major problem. The fuel feed system requires further development to improve reliability.

Further experimental work needs to be carried out on the pilot plant in order to optimise the ODG process and to obtain greater stability in operation before a commercial plant could be designed.

6.0 PROSPECTIVE APPLICATIONS

Factors which will influence the commercial prospects for the ODG process on biomass include capital cost, reliability, start-up/shutdown times and procedures, turndown considerations, the ability (or otherwise) of operators in developing countries to run a less than simple plant, and the performance of the supporting biomass harvesting preparation and supply system.

It is unlikely that 'risk' money will wholly finance the further development work needed to overcome these problems. Further financial support from the EEC, or other governmental agencies is thus essential at the industrial pilot, and demonstration plant stages.

The ODG process for the gasification of biomass has two immediate potential applications :-

- local power generation; or combined heat and power schemes
- industrial fuel gas

At the present stage of development, the process is not suitable for methanol production because the high methane content of the gas would result in poor conversion efficiencies. In addition, the economical uncertanties surrounding this option are such that no commercial case can be made in the short and medium term to invest risk money in its further development. Although the predicted cost of methanol from wood is claimed to be around £170 per tonne, the surplus of present and proposed installed methanol production capacity is such that it is difficult to envisage methanol from biomass being commercially viable before the end of the century unless strategic and foreign exchange considerations in certain developing countries radically alter the present position. At the present time there is no case for carrying out further development work on the ODG process to improve gas quality for methanol production.

Neither John Brown or Wellman intend to pursue this option at this stage.

6.1 LOCAL POWER GENERATION

The ODG process is considered to have considerable potential for local power generation in the range 0.5 - 5 MW in developing countries. Due to the high percentage of sensible heat recovered from the gasification process, the preferred application is combined heat and power schemes where heat is recovered in the form of steam or hot air.

Power generation will be from either a spark ignition gas engine (Otto cycle) or dual fuel engine (Diesel cycle) where the advantages of using a medium calorific value gas over a low calorific value gas are self evident.

These units are of interest in those developing countries that are rich in biomass and where the central electrical transmission system covers a limited area. Such countries include Malaysia, Indonesia, Phillipines, Brazil etc. To be viable, the price has to be around £500/KW for a complete unit and a lot remains to be done before an ODG based unit can be designed and optimised with confidence.

6.2 INDUSTRIAL FUEL GAS

There are two possibilities, the first concerns the retrofitting of large oil and gas fired steam boilers, the second the firing of industrial furnaces and kilns in the metals and ceramics industries.

The determining factor in both these areas will be economics, ie the price at which the fuel gas can be delivered to the burner which is primarily influenced by the price of biomass. The gas produced by an ODG should however be attractive and should not give any problems in its combustion in industrial type furnaces and boilers.

GASIFICATION OF BIOMASS FOR THE PRODUCTION OF SYNTHESIS GAS WITH THE INTENTION TO PRODUCE SYNTHETIC FUEL IN A FURTHER PROCESS

FINAL REPORT

Author	:	G. Cogliati
Location, Country	:	Sulmona, Italy
Contract number	:	ESE-P-070-I (S)
Duration of contract	:	1.4.1982 - 30.6.1985
Total budget of project	:	Lit. 2.073.934.000
CEC contribution	:	Lit. 791.000.000 (38.3%)
Name of other funding bodies and their share of total cost	:	ENEA (Italian Authority for the development of nuclear and renewable energy) - 25%
Head of project	:	Ing. F. Fonzi
Contractors	:	AGIP S.p.A. Via del Serafico, 89 00143 Roma ITALENERGIE S.p.A. Via della Repubblica, 39 67039 Sulmona

Summary

The overall objective of this project is to evaluate the technical and economic feasibility of producing nitrogen free, medium Btu gas from biomass without the use of pure oxygen. The biomass is gasified in a fluidized bed. The heat required for the gasification is supplied indirectly through an outer fluid bed, where a part of the gas produced is burnt with air. Pure oxygen is used only in a separate cracking chamber, if it is deemed necessary, for the production of syngas, to refine the gas by eliminating methane and heavier hydrocarbons.

This paper describes the experimental plant designed in order to test the process indicated above. The main features of the plant are:

- reference feed rate: 500 kg/h of biomass (on dry basis)
- reactor type: fluidized bed with an inert material as fluidizing medium
- heating system: indirect annular chamber around the reaction volume
- gasifier fluid bed temperature: 730°C
- steam/biomass ratio: 0.5
- heat input: 350000kcal/h
- medium-Btu gas produced: 632 kg/h

This gasification process should provide a real improvement over an oxygen blown system in a medium-Btu gas production from biomass and demonstrate the possibility to obtain a product gas with the following characteristics:

- a constant heating value, independent of the feed moisture content, due to the gasification in absence of air;
- a high heating value that allows for its direct use in replacement of oil and natural gas;

TABLE OF CONTENTS

1. INTRODUCTION

1.1. Aim of the project

The increase of the costs of petroleum products, which has occurred throughout the world over the last few years, has goaded industrial operators to look for new energy sources which are easily obtainable, inexpensive, and which use would not create serious environmental problems. Biomass can contribute in solving a number of problems of energy supply. It is, in fact, possible to obtain from it a variety of fuels and chemical products by numerous processes. Biomass represents a source of renewable energy with low sulphur content, and is widely available in many areas of the world.

In greater detail, the ligneous and cellulosic wastes produced by various agricultural and industrial activities constitute energy sources of particular interest because of their chemico-physical characteristics and because of their low cost. We refer, in particular, to timber residues, and waste products of the wine-making, oil-pressing, textile and paper-making industries.

In Italy, the production of wastes from the timber mills and wood-based industries amounts, for example, up to about 1.5 million tonnes a year. These wastes are partly used for local energy production; nevertheless, a large portion remains unused.

Another important energy source consists of presssed olive residues ("olive cake"), which are obtained from olive processing. The average production of this material, which is mostly unutilized, is about one million tonnes a year. Pressed oil residues represent an excellent energy source as they constitute a waste product with relatively low moisture content (15-20%) and a calorific value equivalent to that of wood (about 3,800 kcal/kg). The cost per calory produced by pressed olive residues amounts to about one third of that of oil and to about one half of that of wood.

Lastly, coppice is a presently incompletely used potential biomass resource which can contribute to energy needs.

This report describes the first phase of a project, conducted in Italy with the financial contribution of the Commission of the European Economic Community (Project ESE/070-I) and ENEA (Italian authority for the development of nuclear and renewable energy), during which a fluidized-bed biomass gasification plant, with indirect heating, was constructed.

1.2. The gasification process

The gasification process of wood passes through successive phases which involve drying, pyrolysis, reduction and oxidation procedures. These phases may be outlined as follows:

1st phase - Drying:

Moist wood + heat = dry wood + steam

2nd phase - Pyrolysis:

Dry wood + heat = charcoal (carbon), CO_2, CO, CH_4, organic compound ($C_n H_m$, unsatured and oxygenated compounds).

3rd phase - Gasification:

Carbon + Water + heat = CO + H_2

Carbon + CO_2 = 2 CO

The overall reactions are:

$$C + CO_2 = 2\,CO \qquad \Delta H^\circ = +\,168 \text{ kJ/mole}$$
$$C + H_2O = CO + H_2 \qquad \Delta H^\circ = +\,175 \text{ kJ/mole}$$
$$CO + H_2O = CO_2 + H_2 \qquad \Delta H^\circ = +\,2.9 \text{ kJ/mole}$$
$$C + 2H_2 = CH_4 \qquad \Delta H^\circ = -\,75 \text{ kJ/mole}$$

The fourth reaction is suppressed by operating at a low pressure and at a high temperature.

Partial oxidation with air of the material to be gasified is the solution commonly used to obtain the necessary process heat.

This method makes it possible to use a gasification plant of simple design, but has the disadvantage of producing a gas diluted with the nitrogen from the air. Consequently, the gas produced has a low calorific value (900 - 1.200 kcal/kg). This entails a number of inconveniences: the gasification plant must be directly coupled to the steam generator in order to use the large sensitive heat of the gas produced at a high temperature; the standard steam generator must undergo considerable modification to ensure that high efficiencies are obtained; if the gas is used in internal combustion engines instead of gas oil or petrol, then the gas must be sufficiently purified and cooled.

If it is necessary to produce gas for subsequent use in the synthesis of methanol ("syngas"), or if it is desired to produce a gas of medium calorific value, the gas produced must be free of nitrogen: numerous technical solutions have been proposed in this respect. For example, the process heat can be generated inside the gasification reactor by partial combustion of the biomass with pure oxygen instead of air (1), or with oxygen obtained from the thermal decomposition of particular chemical compounds (2); the process heat, generated outside the gasification reactor, is supplied to the biomass by an inert material circulating between a combustion chamber and the gasifier (3); lastly, the gasifier can be heated externally.

The process we proposed is based on the latter alternative. In comparison to gasification with pure oxygen, our process has the advantage of not requiring an oxygen generation plant for the production of syngas.

The process heat is provided from an annular chamber placed around the gasification reactor; high temperature flue gases, obtained from the combustion with air of a part of the gas produced pass through this chamber. In order to facilitate heat transfer from the flue gases to the gasifier walls, a granular inert material is used in the annular chamber, which is fluidized by the gas. Likewise, the transfer of the heat to the biomass particles is facilitated by a bed of inert particles, fluidized by recycling to the gasifier a part of the gas produced together with some superheated steam.

The composition of the gas produced depends, apart from the reaction temperature, on the rate of heating of the wood particles and on the period during which the reaction products remain at a high temperature. The concentration of methane and of oxygenated and unsatured organic compounds drops if the gasification proceeds at a high temperature and/or long residence times of the reaction products. For economic reasons, and to protect the construction materials, the gasification temperature in our process has been fixed at 730°C. At our operation conditions, the heating time of the wood particles is very short and the average time the reaction products remain at a high temperature is much longer.

Under these conditions the gas contains an elevated percentage of methane and other condensable hydrocarbons, which are undesired if the gas is to be used as a synthesis gas for methanol production. Then it is necessary to heat up the gas in a further step. Therefore a second gasification step is foreseen where partial combustion with oxygen takes place in order to achieve a temperature of 1300°C, which is necessary to crack the hydrocarbons into CO and H_2.

This treatment indeed reduces the advantage of the total independence of oxygen but, nevertheless, the consumption of oxygen is still much less in comparison to the classical oxygen gasification.

This report presents:

- the initial considerations, and preliminary design of the plant;
- the laboratory tests which have been carried out in order to obtain basic figures for the plant design;
- the final design of the plant;
- the results of the test runs carried out with the plant.

2. SMALL SCALE EXPERIMENTS AND PRELIMINARY CALCULATIONS

2.1. Basic calculations

In order to have some figures for the dimensioning of the plant components and for the configuration of the whole plant, the following calculations have been carried out:

- calculation of the heat and the temperature obtained by complete combustion of biomass with air, and composition of the gas produced;
- calculation of the heat consumption for the gasification of biomass without addition of oxygen and steam, and calculation of gas composition at different temperatures;
- calculation of heat comnsumption for the gasification of the biomass at temperatures of 730, 830, 930 and 1,030 °C with different ratios of biomass/steam at steam temperatures of 350°C; calculation of the gas composition and amount of gas produced;

- calculation of the heat consumption for gasification of the biomass with steam and addition of oxygen; calculation of gas composition at different temperatures.

2.2. Preliminary design

The preliminary design of the plant is shown in figure 1.

Biomass with a moisture content of 20% is fed into the fluidized bed with a screw feeder. The gasifier is heated up with heating pipes containing inert fluidizing material. Heat is being produced by combustion of a part of the produced gas and the combustion is taking place inside the pipes.

Gases leaving the gasifier are cleaned from char and ashes by two cyclones, and the separated char is fed again into the gasifier. One part of the cleaned gases is used as flushing gas and is recycled to the gasifier by a hot-gas blower. The temperature in the different sections of the plant is as follows:

fluidized bed of biomass	730°C
fluidized bed of heating pipes	930°C
cracking chamber	1100°C

The gases leaving the cracking chamber have to be cooled in a first stage down to 240°C. After this, fine solid particles like ashes, are separated by a fine mesh filter. In a second cooler the syngas is cooled down to 45°C.

2.3. Laboratory tests

As the basic figures are based on equilibrium calculations, and as there was still uncertainty about the behaviour of the fluidized beds, the heat transfer coefficient between the two fluidized beds and the gasification rates were checked in several tests.

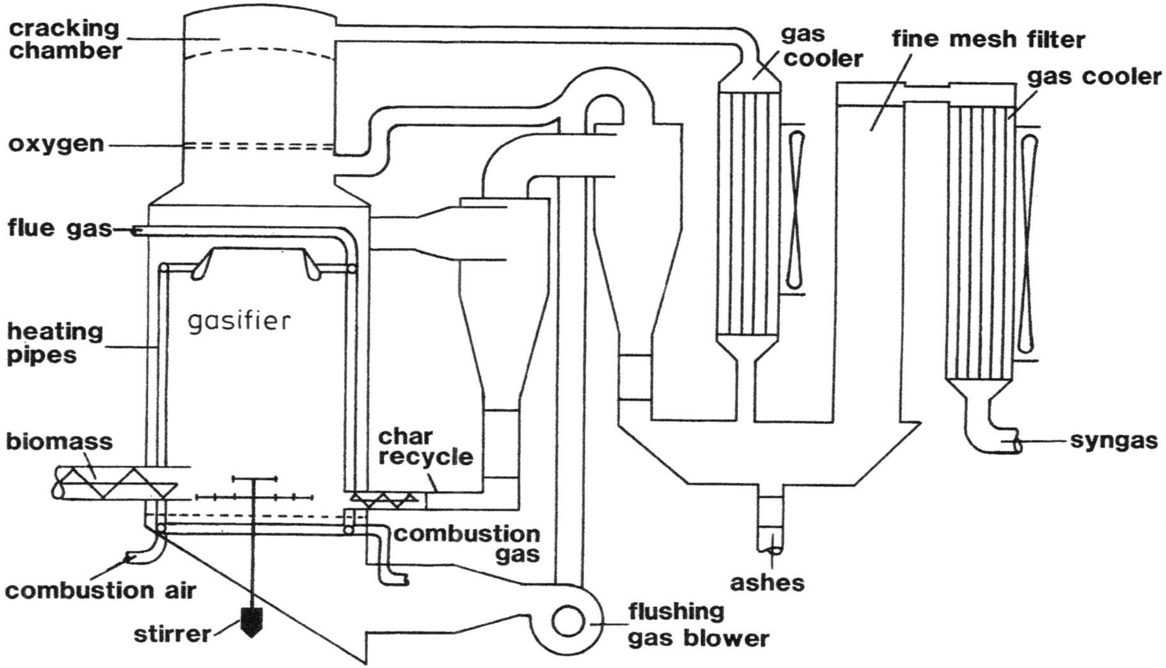

Fig. 1 - Previous scheme of Maremma plant

2.3.1. Test Unit for determination of the heat transfer coefficient

In order to test the heat transfer coefficient between two fluidized beds, a small test unit was constructed, see figure 2.

In this test unit there are two separate fluidized beds, one in an inert tube representing more or less one of the heating pipes of the preliminary design. The outside bed was heated up by the heat produced by burning propane gas in the inner tube.

The results of one of these tests carried out with the unit are shown in table 1

Table 1: results of test run in order to investigate the heat transfer coefficient between two fluidized beds.

air temperature in bottom section T_4	80°C
average air temperature in middle section $T_{1,2,3}$	370°C
air temperature in top section T_5	460°C
flue gas temperature in bottom section T_8	910°C
flue gas temperature in top section T_6	800°C
flue gas temperature in middle section T_7	860°C
air temperature in outlet pipe T_{10}	430°C
pressure drop over outer fluidized bed	1225 Pa
static height of outer and inner fluidized bed	300mm
air flow rate in outer fluidized bed	$351 m^3/h$
average velocity of the air in outer bed	1.78m/sec
coefficient of heat transfer k	$303 W/m^2K$

The numbers of the measuring points correspond with the numbers in Figure 2.

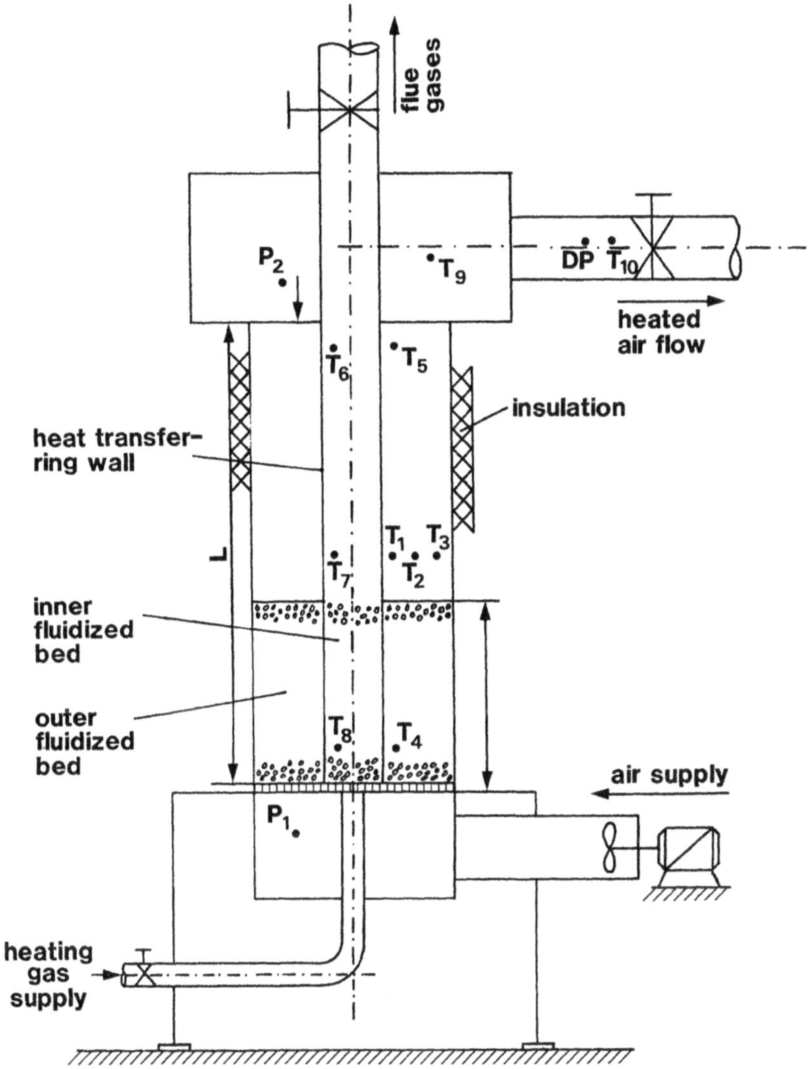

Fig. 2 - Scheme of test unit

After completion of the heat transfer tests, the plant was used to carry out further test runs in which the following parameters were investigated and/or optimized:

- testing prospective fluidizing materials with respect to heat resistance, abrasion, specific weight, particle size and abrasion of metallic heating surface;
- realization of a smooth fluidization with no slugging and a coefficient of heat transfer above $250 W/m^2K$;
- investigation of the minimum fluidization velocity, the turn-down ratio and the solids entrainment characteristics.

In order to carry out these tests the plant was modified several times. For example, for testing the fluidizing behaviour of the biomass, the inner tube was removed and the biomass was fluidized with air at ambient temperature.
The results and the conclusions of all these tests are as follows:

From the various fluidizing materials tested: ceramic balls, quartz sand and Al_2O_3-balls, the latter turned out to be the best. As in these test runs it was not possible to realize the high temperature of 1,200°C which will occur in the gasifier, the balls could not be tested for abrasion at such high temperatures. A special burner and combustion chamber was realized for that purpose.
Fluidization in tubes of relatively small diameters tend to generate undesired pulsations. In order to avoid this effect, several internals such as a spiral and a set of sieves were tested; but without the desired results: the use of heating tubes appeared to remain unreliable and therefore this concept was dropped. Alternatively, it was decided to transfer heat from an outer heating chamber.

2.3.2. Laboratory test on gasification

In order to calculate the volume of the gasifier, the residence time required for complete gasification of biomass, has to be known. Therefore laboratory tests were carried out with an arrangement as shown in Figure 3.

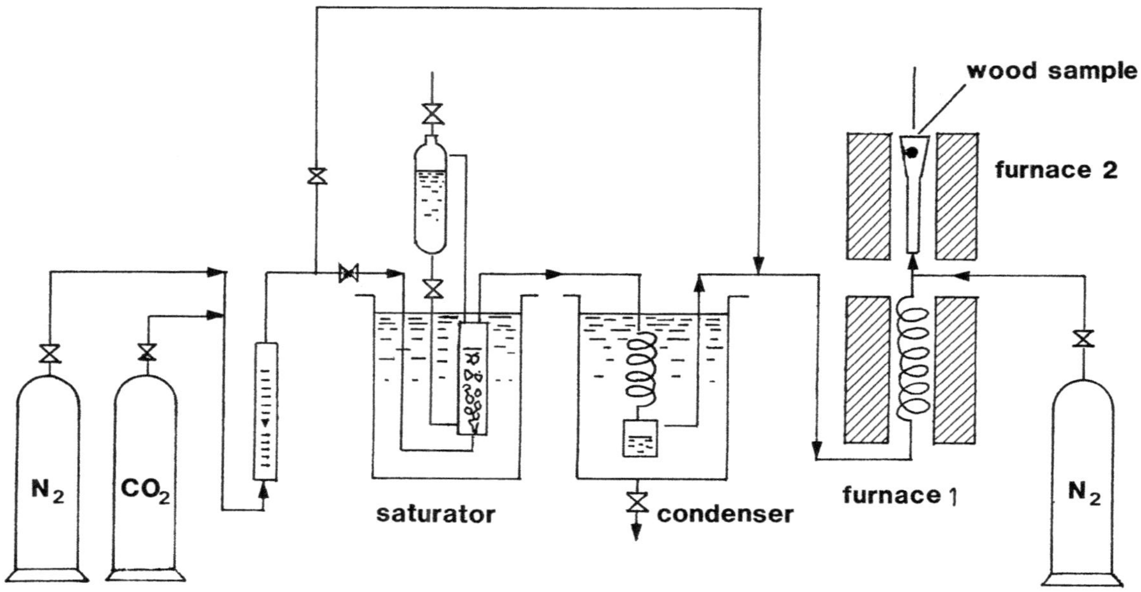

Fig. 3 - Laboratory arrangement for determination of gasification time

This arrangement makes it possible to produce a gas composition consisting of CO_2, N_2 and H_2O in any percentage. N_2 and CO_2 coming from pressure cylinders were either directly heated up in furnace no. 1 or first saturated with steam. When the gases were heated up to the desired temperature, the wood sample, placed in a platinum container, was introduced into the second furnace: this sample was removed after a given time and immediately cooled with inert gas. Then, the weight loss was determined.

It has been ascertained that the weight loss rate is very large initially. This is clearly shown in Figure 4, which reports the weight loss of a fragment of dry wood placed in a nitrogen atmosphere at temperatures ranging between 650 °C and 850 °C: after less than one minute, the pyrolysis is completed where after the weight of the sample remains practically constant: it accounts for 15-18% of the initial weight.

If the heating gas is carbon dioxide, or if steam is added, the weight loss of the sample is greater and the final residue consists of ashes only. Then the wood is completely gasified since any remaining carbon reacts with CO_2 and steam, under further formation of CO and hydrogen (figs. 5 and 6). The kinetics of gasification depend on both the temperature of the reaction and the H_2O/gas ratio (as shown in figs. 7 and 8).

The main results of these tests are as follows:

- At temperatures above 600°C the volatiles (80% by weight) are produced and emitted within one minute);
- The gasification of the remaining charcoal is improved by the use of steam. Therefore, for the pilot plant, it is foreseen to supply 0.5 kg steam per kg of biomass;
- The remaining charcoal is almost completely gasified in 20 minutes under conditions as foreseen for the pilot plant (730°C, absence of oxygen, steam rate = 0.5 biomass rate).

2.3.3. Conclusions from the laboratory tests

The test runs for investigation of the heat transfer and the gasification time have resulted in the following conclusions:

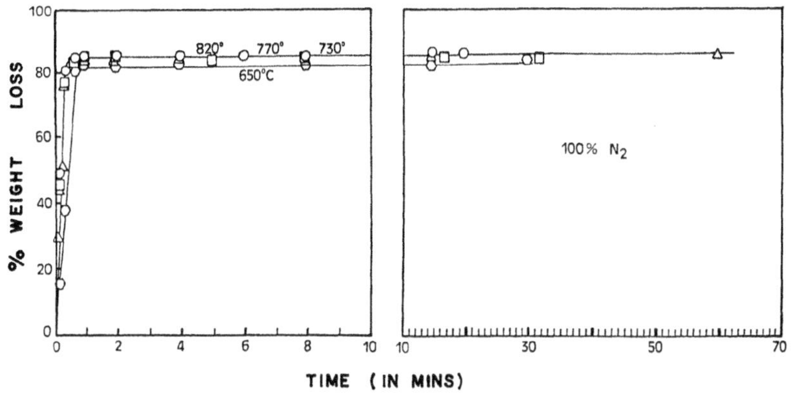

Fig. 4 Pyrolysis of wood chips in atmosphere of N_2

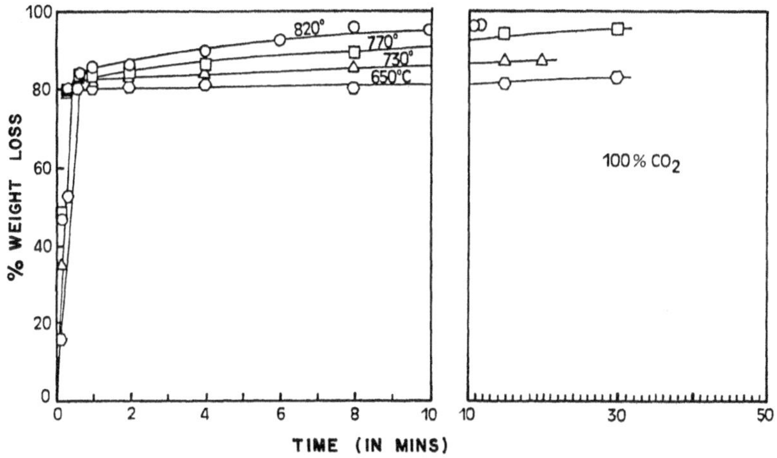

Fig. 5 Pyrolysis/gasification of wood chips in atmosphere of CO_2

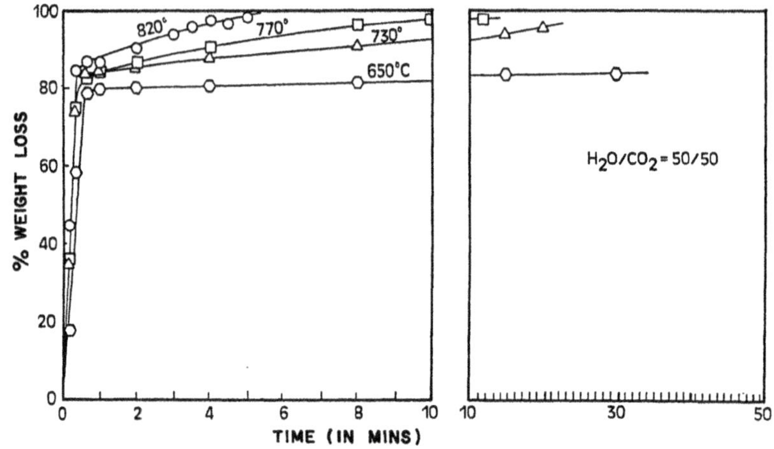

Fig. 6 Pyrolysis/gasification of wood chips in atmosphere of H_2O-CO_2 mixture

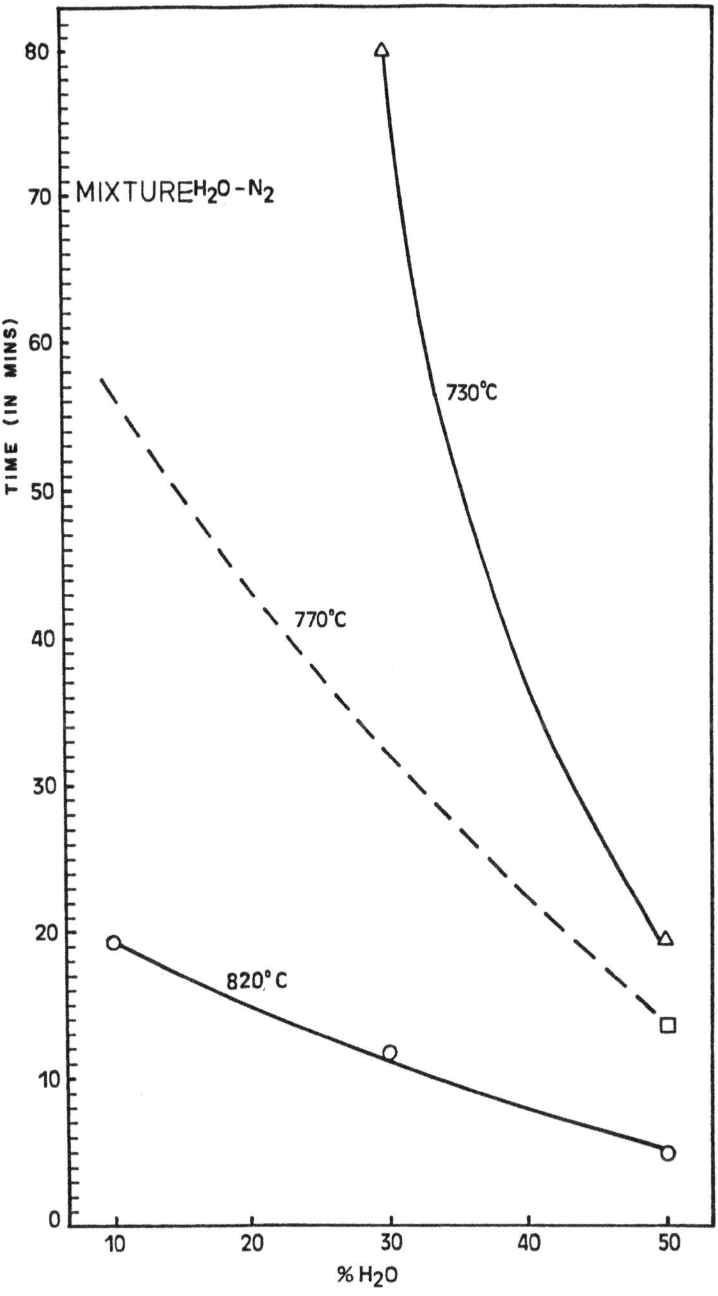

Fig. 7 Contact time needed to reach a 97% weight loss in terms of the content in the H_2O-N_2 mixture

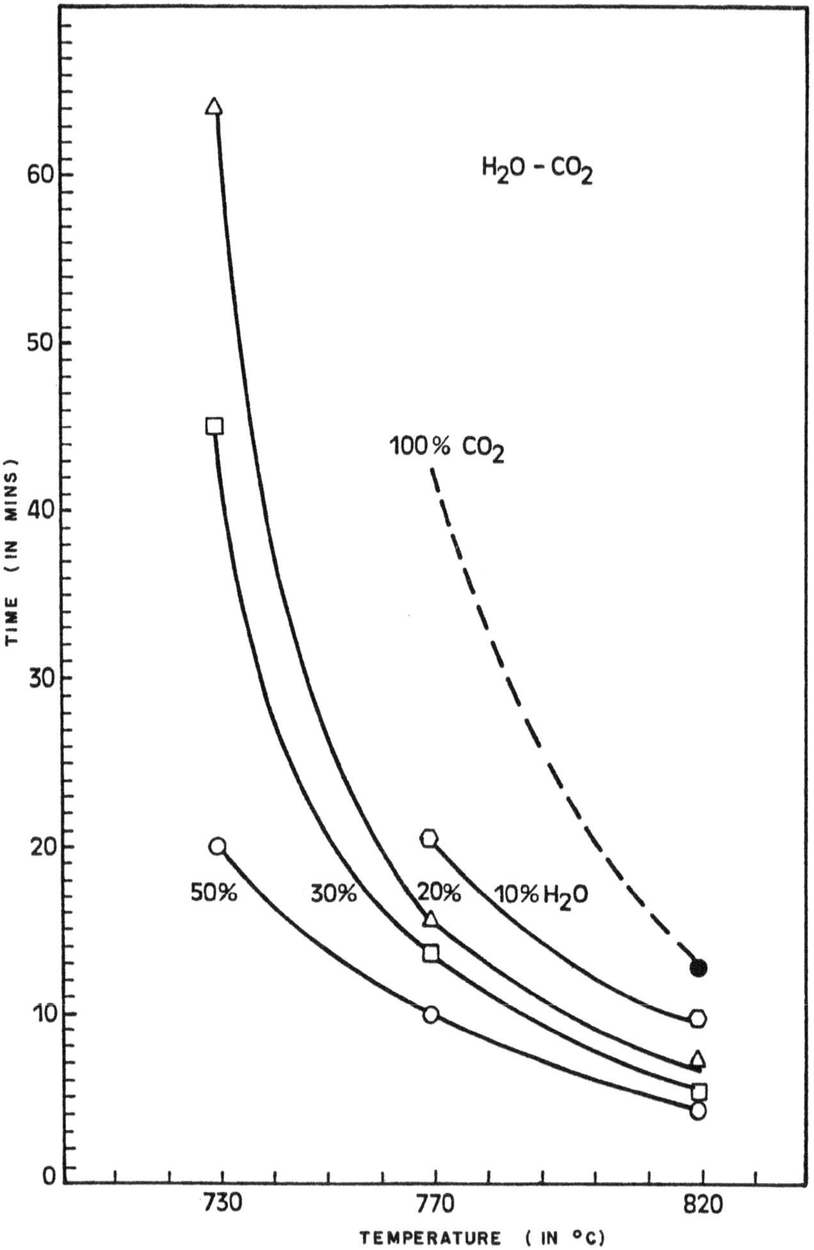

Fig. 8 Contact time needed to reach a 97% weight loss as function of the reaction temperature for various compositions of the H_2O-CO_2 mixture

- A heat transfer coefficient of 260 kcal/h.m^2.K has been observed between the two fluidized beds. As the conditions for the test run were not exactly the same as in the pilot plant (higher temperature difference between the two beds), a value of only 190 kcal/h.m^2.K has been assumed for the pilot plant.
- under steady state process conditions, complete gasification takes place in 20 minutes, while the escape of the volatiles produced by pyrolysis is already completed after 1 minute and amounts up to 80% of the biomass mass. It is assumed that 20% of the biomass mass feed rate x 20 minutes gives the amount of charcoal always present in the gasifier.

3. DESCRIPTION OF THE PLANT (The process flow diagram of the plant is shown in fig. 9)

3.1. Biomass feeding system (X-01)

The main objective of the project is the gasification of biomass. The equipment for preparation of the chips and for chips drying up to a fixed moisture-content of 20% of water by weight, in comparison with dry wood, are standard items which have not been included in this project.

The biomass feeding system consists of the following main items:

- A biomass storage facility with a capacity of two days' operation of the gasifier (30 tonnes of biomass - volume 80 m^3);
- A biomass transfer and weighing system, consisting of conveyor belts, elevators, granular materials scale, separator of ferrous foreign materials, hoppers, etc. in order to guarantee the feeding of the wood chips to the gasifier at a variable rate from 100 to 1,000 kg/hour.

3.2. System for producing gasifier heating fluid

As previously stated, the reactions that occur in gasification of biomass are endothermic: the heat is supplied to the system by means of a fluid which heats the outer walls of the reactor. This fluid consists of flue gases obtained from combustion of a part of the gas produced in the gasifier with excess air.

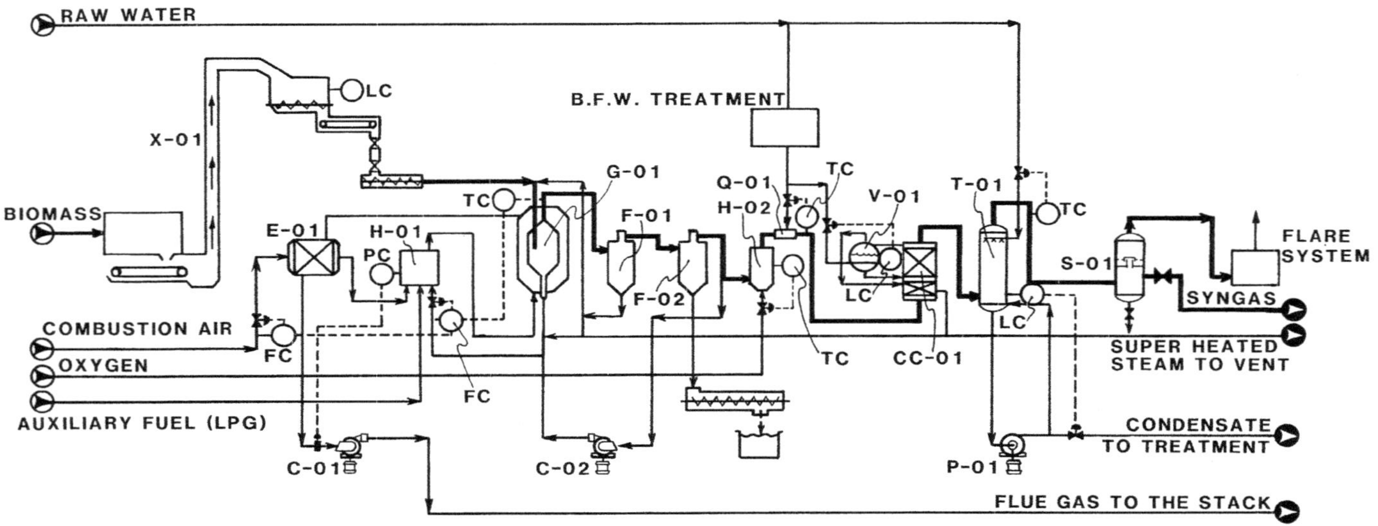

X-01: LOADING SYSTEM
E-01: AIR PRE-HEATER 355000 KCAL/H
C-01: I.D. FAN
H-01: COMBUSTION CHAMBER 506000 KCAL/H
G-01: GASIFIER 301000 KCAL/H
C-02: F.D. FAN
F-01: PRIMARY CYCLONE
F-02: SECONDARY CYCLONE
H-02: CRACKING CHAMBER
Q-01: QUENCHER
V-01: STEAM DRUM
CC-01: STEAM GENERATION E SSH COIL 265000 KCAL/H
T-01: SPRAY CHAMBER
P-01: CONDENSATE PUMP
S-01: KO DRUM
X-02: FLARE SYSTEM

Fig. 9 - Process flow diagram of the gasification plant

As will be described under 3.3, the amount of heat to be supplied to the reactor, under the assumed operating conditions, amounts to 300,000 kcal/hour. Because of the behaviour of the materials, the maximum temperature of the flue gases is fixed at 1,250°C; the temperature difference between the entry and the exit of these flue gases from the reactor is fixed at 400°C. Under these conditions, the rate of flow of the flue gases amounts up to 2,500kg/hour. The air fed to the combustion chamber is pre-heated in a heat-exchanger by the flue gases coming from the gasifier, which cool down from 850°C to 340°C. In this experimental plant, these flue gases are no longer used, and are discharged directly into the chimney stack.

It is obvious that in an industrial plant, the residual heat would be recovered and used, for instance, for drying the biomass or for generating electric power to serve the plant.

Upon starting up of the plant, when the production of gas from the biomass has not yet started, or is still insufficient, the production of the heating fluid is assured by auxiliary fuel.

A fan, mounted downstream of both the combustion chamber and the preheater, ensures the desired flow rate of hot flue gases through the gasifier heating chamber.

Adequate instrumentation controls both the maximum temperature of the flue gases, fixed at 1,250°C, and the gasification temperature of 730°C by adjustment of the air and the combustible gas flow rates, respectively.

The combustion chamber is shown in fig. 10.

3.3. Gasifier

The gasifier is shown in fig. 11

The system consists of a reactor, cyclones for the separation of the solid products from the gas, and a compressor for recycling the fluidizing gas.

The gasification of biomass, by heating up and by subsequent reaction with steam, occurs in the inner chamber of the reactor. On fixing the hourly flow rate of the biomass (500 kg/h) and the steam biomass ratio (0.4), the calculated hourly quantity of heat that has to be supplied amounts up to about 200,000 kcal/hour. In order to allow for unavoidable heat losses through the surface of the equipment, and to ensure an adequate safety margin, the quantity of

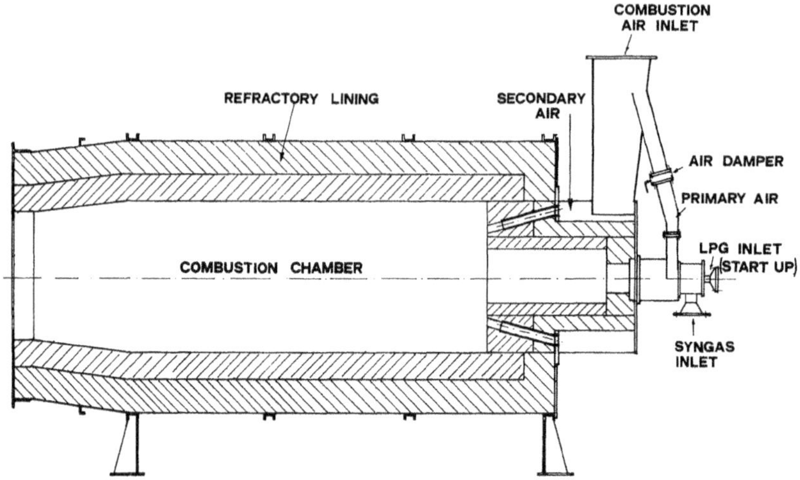

Fig. 10 The combustion chamber

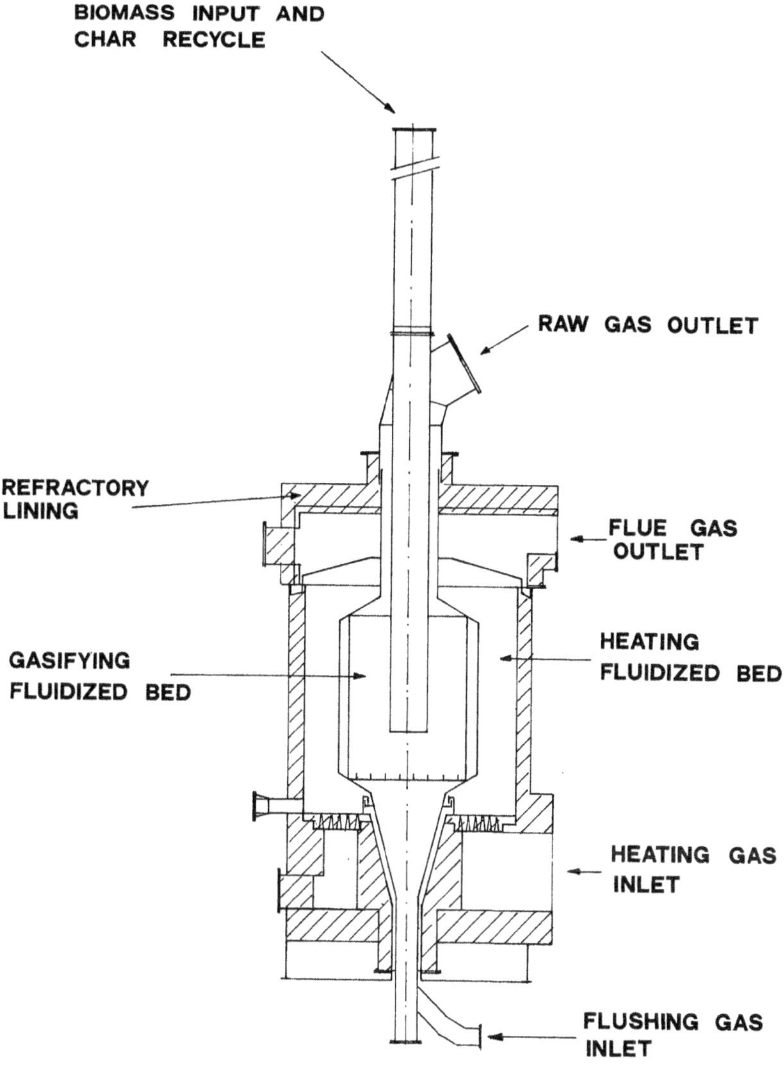

Fig. 11 Fluid-bed gasifier

heat to be provided to the gasifier has been estimated at 300,000 kcal/hour.

As has been stated in 4.2, the temperature difference between the entry and the exit of the heating chamber has been fixed at 400°C. Assuming further that, under processing conditions, the heat-exchange coefficient through the separating walls of the two chambers amounts to 190 kcal/m^2°C, the exchange area necessary to transfer 300,000 kcal/h amounts to about 4 m^2, which must be completely covered by fluid beds at both sides. The fan, which recirculates the gas to the gasification chamber, operates under particularly difficult conditions (730°C) and unless highly sophisticated equipment is supplied, its head is limited to about 800 mm of water. To obtain fluidifisation, it is necessary to overcome the hydrostatic head of the solid. As the discharge head of the fan is rather low, it is necessary to limit the height of the bed and/or the density of the particles.

To realize a low pressure drop, the side wall of the gasifying chamber has been constructed in corrugated sheet: the necessary exchange area is thus obtained with a smaller height of the gasification chamber. Considering an average diameter of 1,200 mm for the gasification chamber; the exchange area of 4 m^2 is obtained with a 700 mm high fluidized bed. If the discharge head of the fan is 800 mm of water, the density of the bed under steady state conditions must be less than 1.15 kg/dm^3. In practice, allowing for the pressure losses due to the diffuser and to the pipe circuit, the apparent density of the bed must not exceed 0.8 - 0.9 kg/dm^3.

Therefore, we selected as an inert material for the fluid bed, 1 - 2 mm diameters feroidal particles of alumina with a bulk density of 0.8 kg/dm^3. Laboratory tests have shown that this material has a good mechanical strength and a modest erosion effect on the metal walls.

The fluidization velocity under operation will be about 1.5 times the minimum fluidization velocity calculated on the basis of Ergun's well known formula for 1-2 mm alumina particles. The fluidization velocity is 0.5 m/sec. Under these conditions a good bed turbulence is obtained without causing entrainment of either the finer alumina particles, or the coarser char particles. Only the ash particles will be entrained.

It is difficult to predict the behaviour of the biomass in the fluid bed of alumina particles under the operating conditions. Since the biomass is fed from the top, it is likely that the mixing of the wood chips with the alumina particles would be limited. If so, two overlapping beds would be obtained, one of alumina particles and one of wood chips, the latter merely heated by the gas flow which in turn is heated up upon passing through the inert fluidized bed.

Since the volume of the gasification chamber is 1,700 litres, 850 litres of which are taken by the inert material, the biomass (600 kg/hour, apparent density 300 kg/m^3, equivalent to 2 m^3/hr) has an available reaction volume of about 850 litres which allows for a gasification time longer than the experimental one of 20 min., determined by the laboratory tests under the same temperature conditions (730°C).

Therefore the wood particles should be pyrolysed in less than a minute into char and subsequently be gasified by CO_2 and/or steam, present in the fluidising gas. Only particles which have been completely gasified, that is which size has been reduced down to the micron scale, may leave the gasifier by entrainment. Therefore the first cyclone is designed to separate from the gas all solid particles with a grain-size over 500 um; these particles still contain char, and they are, consequently, recycled to the gasification chamber.

Two further cyclones separate the finer particles, which are considered to consist mainly of "ash": these particles are discharged from the plant and collected in an appropriate container. Naturally, these "ashes" will contain a part of fine-grained char, which has not reacted. The flow rate of the ashes has been estimated at 3.8% of the biomass introduced, i.e. 23 kg/hour.

The char separated in the first cyclone is recycled to the gasification chamber, using the biomass feed channel. The solids downflow is driven by gravity. The cyclone is separated from the gasification chamber by two valves which open alternately: the chamber acts as an intermediate container between the two valves. The feeding of the char into the gasifier is facilitated by supplying a part of the superheated process steam to the outlet pipe.

The outside chamber of the reactor is cylindrical, with metal side walls lined with insulating and refractory material; the flue gases are fed through the annular bottom which is made of a drilled ceramic plate. The fluidized bed in this chamber consists of alumina particles, 1.2 mm in diameter. The velocity of the flue gases is typically 1.4 m/sec., i.e., much greater than the gas velocity in the gasification chamber. As is known, a high fluidifisation velocity increases the turbulence of the bed and leads, within certain limits, to a greater heat exchange.

3.4. Cracking chamber

The cracking chamber consists of a metal cylinder lined internally by refractory material (fig. 12).

A suitable burner brings the raw synthesis gas into contact with oxygen which results in partial combustion of the gas. The temperature reached in the chamber can be controlled by varying the oxygen flow rate. The residence time of the gas at the reference temperature of 1,300°C is about one second.

Immediately after leaving the cracking chamber, the gases are cooled down to a temperature of 840°C by injection of water. The operation is necessary to permit the use of a conventional technology for the construction of the steam generator to feed the gasifier.

The gas temperature at the cooling chamber outlet is controlled by varying the water flow-rate.

In the quencher, (fig. 13) an injector in Incoloy 800 is installed through which water is sprayed. The refractory zone, included in the space of about one metre from the injector, is also lined with Incoloy 800 sheeting. The position where the water is injected has an optimal configuration with respect to phase contacting and mixing, between gas and water.

3.5. Super-heated steam generating system. Gas cooling system

After cooling, the gas is delivered to the steam generation system designed to obtain 240 kg/h of superheated steam at a temperature of 600°C and at a pressure of 6 bar. Most of the steam is used for the gasification, while the remaining part is used in the biomass feed system to the gasifier, and in the ash discharge system.

The gas leaves the steam generation system at a temperature of 180°C. Subsequent cooling to ambient temperature follows in a spray tower where water is injected (fig. 14). Here, a good contact is accomplished by a set of contact plates. The gas passes through the plate holes and the water is in cross flow with the gas. The equipment consists of three sections: pre-washing and saturation, water removal, and separation of the drops. The drops, entrained by the gas, leave the upper plate and are then separated in the successive inertial-type separator. Finally, the gas is delivered to the flare system.

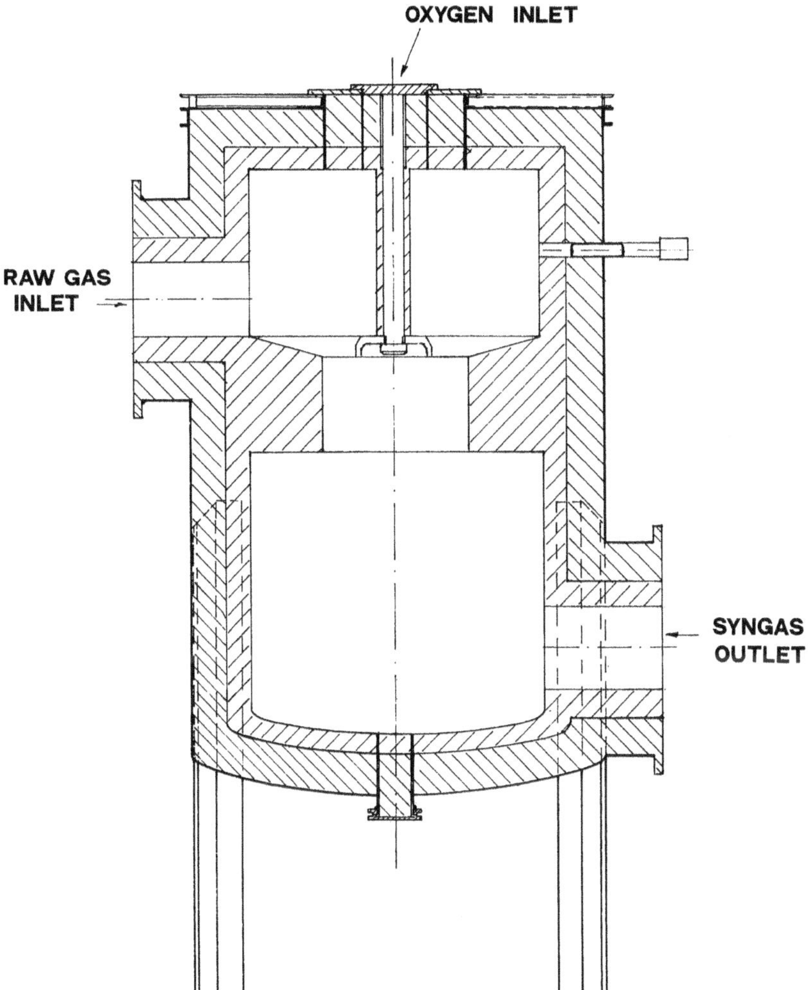

Fig. 12 - The cracking chamber

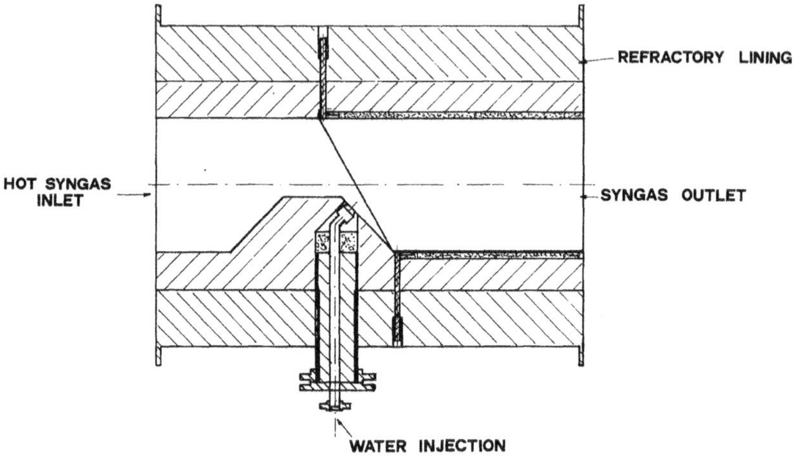

Fig. 13 Water quencher

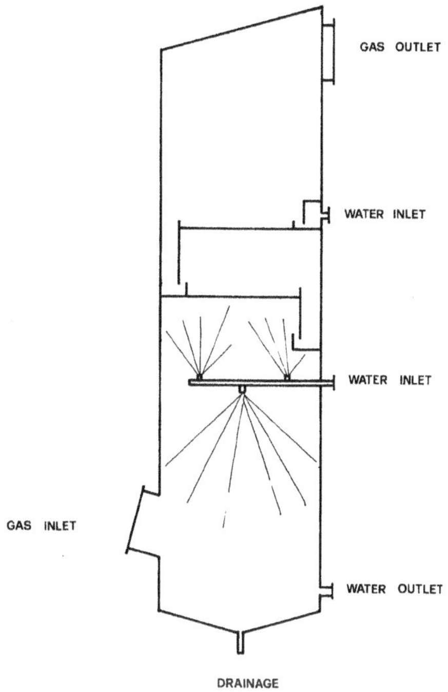

Fig. 14 The spray tower

4. P. & I. - DIAGRAM

The P & I - diagram of the plant is shown in the appendix. Sheet n. 1 represents the gasifier with the cyclones and the recycling gas fan; sheet n. 2 gives the cracking chamber, quencher, steam generator and superheater, spray tower and k.o. drum; sheet n. 3 gives the heat exchanger, the combustion chamber and the flare.

The control system of the plant based in this diagram has the following controlling and shut down functions.

4.1. Control functions

Gasification temperature; this temperature is controlled by the flow rate of combustion air to the combustion chamber.

Temperature of the cracking chamber: controlled by oxygen feed to the cracking chamber.

Gas temperature in the quencher: controlled by the water input to the quencher.

Gas temperature when leaving the spray tower: controlled by the water input to the spray tower.

4.2. Shut-down functions

The following events cause a general shut down of the plant:

- exceeding set temperature of heating gas;
- exceeding set pressure in gasifier (internal bed);
- shut down of flushing gas tank;
- lowest level in steam drum;
- too low input of combustion air;
- too low pressure of lpg at battery limit;
- flame failure in combustion chamber;
- exceeding set temperature in combustion chamber;
- flame failure in flare system.

In all these cases the general shut down is effected in the following way:

- shut down of biomass input into gasifier;
- shut down of gas burner in combustion chamber;
- shut down of oxygen input into cracking chamber;
- shut down of char recycle into gasifier.

Furthermore the control system of the plant effects the following partial shut-down functions:

- exceeding temperature in cracking chamber;
- shut down of oxygen input into cracking chamber;
- opening of emergency water input into cracking chamber;
- exceeding temperature of syngas quencher;
- opening of emergency water input into quencher;
- shut down of oxygen input into cracking chamber;
- shut down of condensate pump:
  - opening of by-pass valve of condensate pump;
- high level in spray tower:
  - opening of by-pass valve of condensate pump.

4.3. Alarm functions

In the following cases an acoustic alarm is triggered

- temperature in combustion chamber exceeds 1300°C;
- too low input of lpg to combustion chamber;
- temperature in gasifier (internal bed) exceeds 780°C;
- too low input of syngas to combustion chamber;
- temperature in gasifier (internal bed) falls below 680°C;
- too low input of combustion air to combustion chamber;
- flue gas temperature at gasifier outlet exceeds 950°C;
- too low input of flushing gas to gasifier;
- temperature of syngas after quencher exceeds 890°C;
- excess pressure in gasifier falls below 400 mm. H_2O;
- temperature in cracking chamber exceeds 1350°C.

5. MASS AND ENERGY BALANCES

Fig. 15 shows the calculated mass balance of the plant.

By feeding 600 kg/h of wood chips containing 20% of moisture by weight and 240 kg/h of superheated steam to the gasifier, 827 kg/h of gas is obtained of which 185 kg/h is used to obtain the process heat. In addition 23 kg/h of ashes are produced. According to estimates based on the chemical equilibrium at 750°C, the gas produced has the following composition:

H_2	(% by volume)	= 32	H_2O	(% by volume)	= 23
CO	"	= 22	CH_4	"	= 8
CO_2	"	= 15			

The calorific value of this gas is 2,700 kcal/kg: it may, therefore, be considered as gas of medium calorific value and, as such, can be used directly for the generation of steam and/or electric power in substitution of natural gas.

The addition of 98 kg/h of oxygen and the resulting temperature increase of the gas in the cracking chamber permits the almost quantitative completion of the reaction:

$$CH_4 + H_2O = CO + 3\ H_2$$

In this way, the reduction of the methane content in the gas produced is accomplished, and 730 kg/h of syngas with the following composition based on thermodynamic equilibrium condition at 1200°C is obtained

H_2	(% by volume)	= 31	H_2O	(% by volume)	= 30
CO	"	= 24	CH_4	"	= 1
CO_2	"	= 14			

Downstream cooling of the gas by direct injection of water, results in condensation of most of the steam present in the gas; the remaining 543 kg/h of gas has the following composition:

H_2	(% by volume)	= 44	H_2O	(% by volume)	= 4
CO	"	= 34	CH_4	"	= 2
CO_2	"	= 16			

This gas should meet the specifications for the synthesis of methanol. Fig. 16 shows the energy balance calculated for the gasification plant. This balance shows that from the 1.98 Gcal/h introduced with the biomass into the gasifier, 1.47 Gcal/h is obtained from the system in the form of syngas. This corresponds to an efficiency calculated at 74%.

This balance does not include the thermal leakages through the surfaces of the equipment and the energy consumption of the pumps, the fans, the instruments, etc. It has, however, to be noted that, since we are dealing with an experimental plant, the possible energy recoveries have not been optimized: for example, the sensitive heat of the flue gas discharged into the stack at a temperature of 340°C can be used for drying the biomass.

6. ENVIRONMENTAL AND SAFETY IMPACTS

The gasification plant, because of its modest capacity, does not involve serious problems regarding safety and environmental impact.

The need to prevent the development and ignition of an explosive mixture in the gasifier during the first few minutes of operation, during shut down, and during emergencies has been considered. Alarms and automatic equipments, are installed on the gasifier to accomplish a safe shutdown in the event of any of the major operating parameters getting beyond the set-values. Modification work, maintenance work, and cleaning of the gasifier are all carried out in accordance with the recommended procedures. The health hazards related to the ash removal are very modest in view of the small quantities involved and this is why it is not considered necessary to install a closed and automatic system. The ashes are, however, kept moist to minimize the hazards of ignition.

The major hazard for the workers is constituted by accidental leaks of the gas produced: this gas, in fact, contains potentially dangerous substances like hydrogen, carbon monoxide and methane.

Even though the plant is in the open air, a continuous monitor for carbon monoxide is always in operation.

General fire protection measurements have been taken in accordance with the existing regulations and fire brigade plans. Portable fire extinguishers are installed at

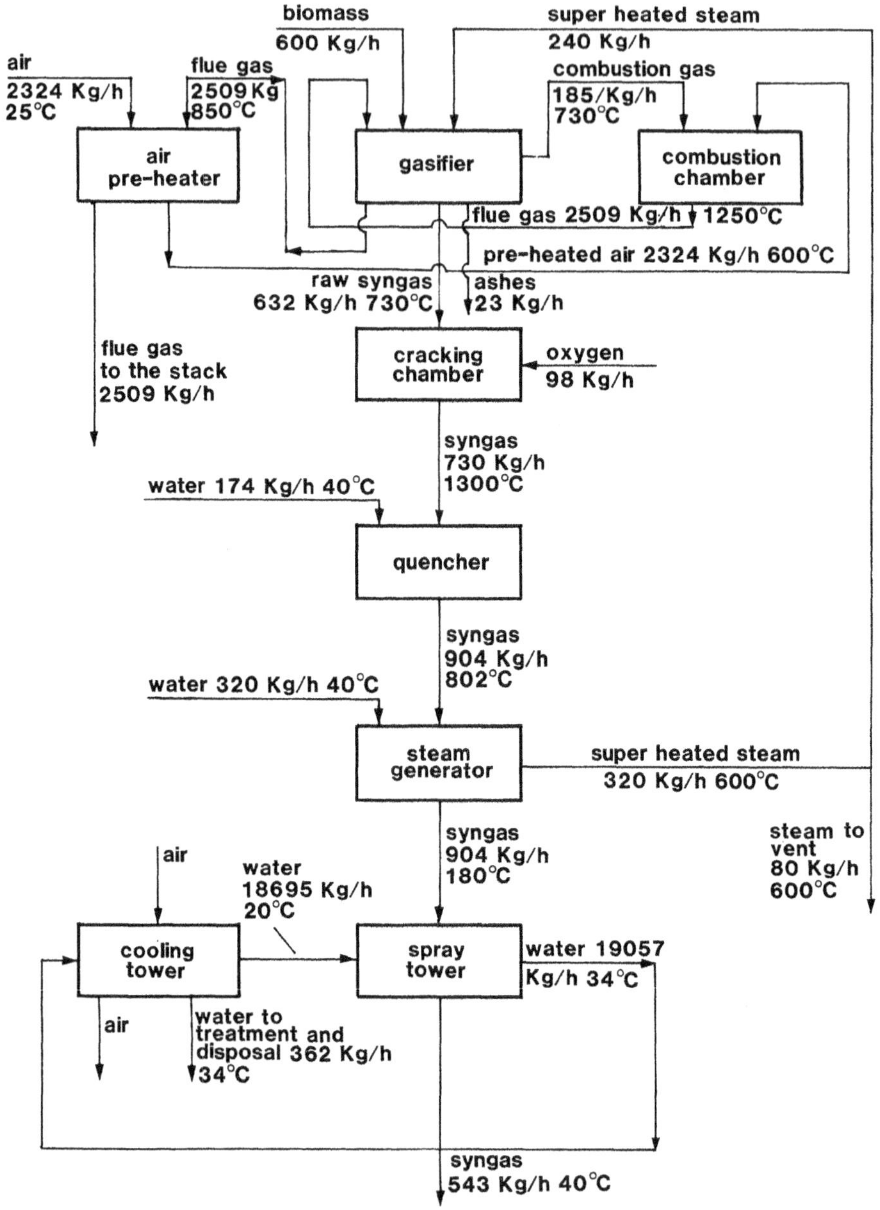

Fig. 15. - Calculated mass balance of the gasification plant

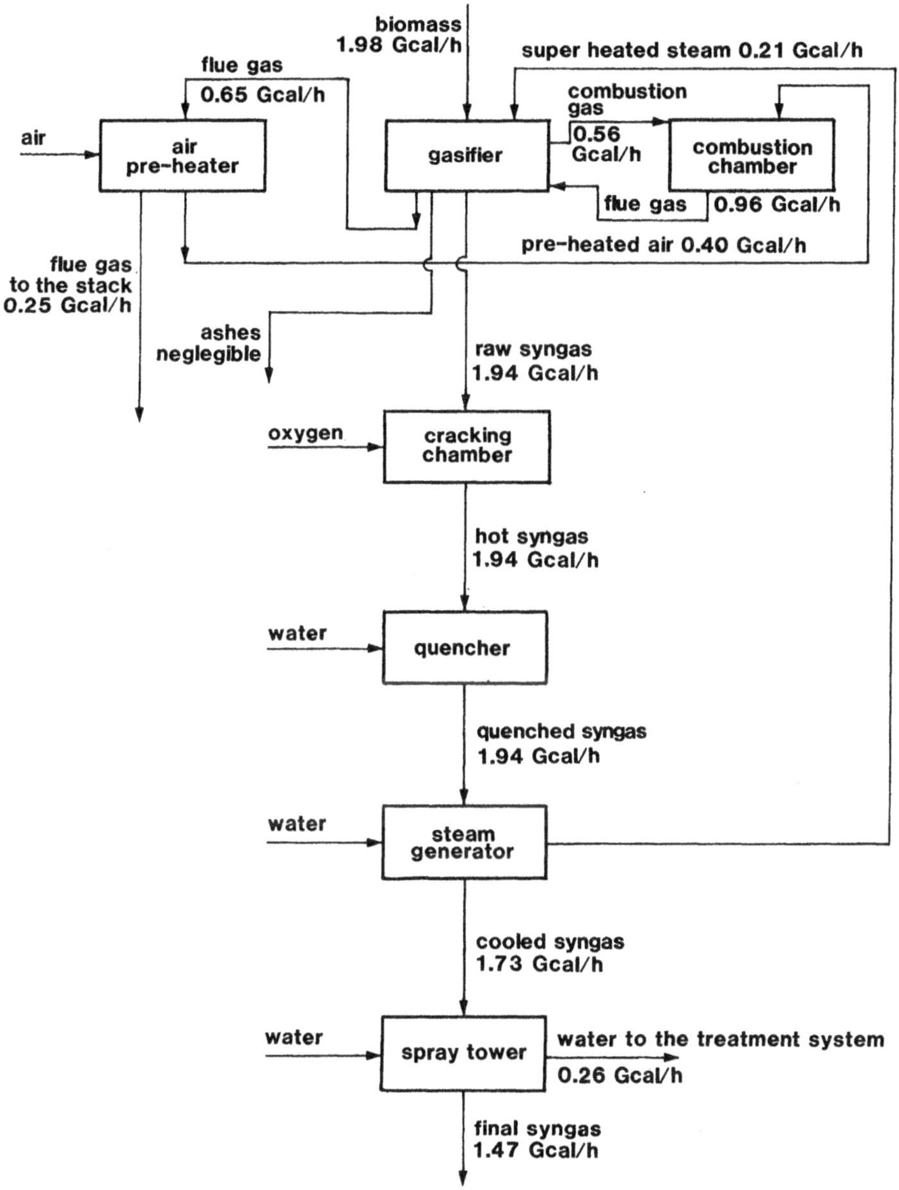

Fig. 16 - Calculated energy balance of the gasification plant marked locations in the sheltered areas, and fixed stations are installed along the outside perimeter of the process area.

Insulation or thermal shields are installed around process components that could cause heat hazards to employees. Asbertos-free insulating material is used.

Each employee has been informed of the potential safety and health hazards of the process and is familiar with the location of all applicable emergency equipment and supplies, and is familiar with its application.
A continuing education program is maintained to keep employees educated in job hazards, proper maintenance procedures, clean up procedures, use of personal protective equipment, etc.

As to the environmental impact, it is recalled that the plant wastes consist of flue gases from the combustion chamber, discharged into the atmosphere through the chimney stack, and of water obtained from the spray-tower cooling of the gas produced.

As to the flue gases discharged through the chimney stack, their SO_2 and NO_2 is especially low because the biomass contains only traces of sulphur and nitrogen, and the gasification is conducted without air.

The low production of SO_2 and NO_2 is a major environmental advantage of this process for wood gasification.

Also the particulate matter content of these fumes should be very low since they are produced by gases combustion at very high temperature (1300°C).

The water from the gas cooling tower will contain only a negligible quantity of char and condensates, since the gas has undergone a cracking treatment at a high temperature which has eliminated all the organic products of high molecular weight present in the gas at the gasifier outlet.

The treatment of this water, therefore, can be limited to filtering.

7. COMMISSIONING OF THE PLANT

The following tests have been carried out:

- pipes cleaning
- tightness tests of equipment and pipes
- calibration of controlling instruments
- inert material charging (1-1.5 mm diameter spheroidal hollow alumina particles) in the heating and gasification chamber
- refractory drying by gradual heating, utilizing the flue gases produced by combustion of lpg inside the combustion chamber.

Design temperatures have been obtained in the combustion chamber (1200°C), in the heating chamber of the gasifier (840°C) and in the inner gasification chamber (730°C). Nevertheless, since the insulating material around the pipes is not yet installed, temperatures lower than the design ones have been obtained in the cyclones and inside the cracking chamber. Subsequent test runs of the plant have shown an irregular behaviour of the pressures and of the flow of the flue gases circulating in the heating chamber of the gasifier and of the air circulating in the gasification chamber. It was noticed that this was caused by damage to the tightness joints of the gasification chamber and the resulting leakage of the syngas. Therefore, it was necessary to replace the existing sintered clay fibre joints by all metal joints. In addition, extra differential manometers were installed in order to measure pressure drops over the heating chamber. Hollow alumina particles were replaced by 1.5 - 2 mm porous alumina particles in order to limit the pulverization of the inert material due to abrasion.

Further tests resulted in the breaking of the refractory heating chamber diffuser. This has been ascribed to the accumulation of alumina dust in the thermal expansion device of the gasification chamber.

After repairing, gasification tests have been started under the following conditions:

Heating circuit:

- Flow rate of combustion air : 1100 kg/h
- Combustion air temperature after pre-heating : 520°C
- Flow rate of lpg 30 kg/h

- Flue gases temperature at the exit of the combustion chamber	: 1200°C
- Average temperature inside the heating chamber	: 840°C
- Flue gases temperature to chimney stack	: 340°C
- Δp of the diffuser	: 200mm wc
- Δp of the fluidized bed	: 600mm wc

Syngas circuit:

- Flow rate of flushing gas	: 2500 m^3/h
- Temperature of flushing gas fed to the gasifier	: 170°C
- Syngas temperature at the exit of the gasifier	: 700°C
- Syngas pressure	: 300mm wc
- Flow rate of biomass	: 200 kg/h
- Flow rate of syngas to the torch	not measured

Average analysis of the produced gas:

H_2	(% by volume)	20.5
CO	(")	18.8
CO_2	(")	> 20
CH_4	(")	7.1

The gasification tests, carried out up till now must be considered as preliminary, since biomass has been charged at a flow rate of around 1/3 compared to the design value (200 kg/h instead of 600 kg/h of biomass), and for too short periods of time to achieve equilibrium conditions.

Nevertheless, the innovating concept on which the gasification plant is based (indirect heating by fluidized beds) seems to be proven: the possibility of obtaining a medium Btu gas without employing pure oxygen has been demonstrated.

Therefore, now it is necessary to attain the demonstration of the technical and economical validity of the process through a series of experiments under the working conditions settled in the project. In addition to wood chips, other kinds of biomass, such as RDF, will be applied as well.

8. REFERENCES

1. Chartier, P. and Palz, W., Solar Energy R&D, in the European Community, Vol. 1, 198, Reidel, Dordrecht (1982).

2. US Patent 7900761, 9th January 1979; Eur Patent 13590.

3. Feldmann, H.F., Energy from Biomass and Wastes V, IGT (1981).

SHEET 1

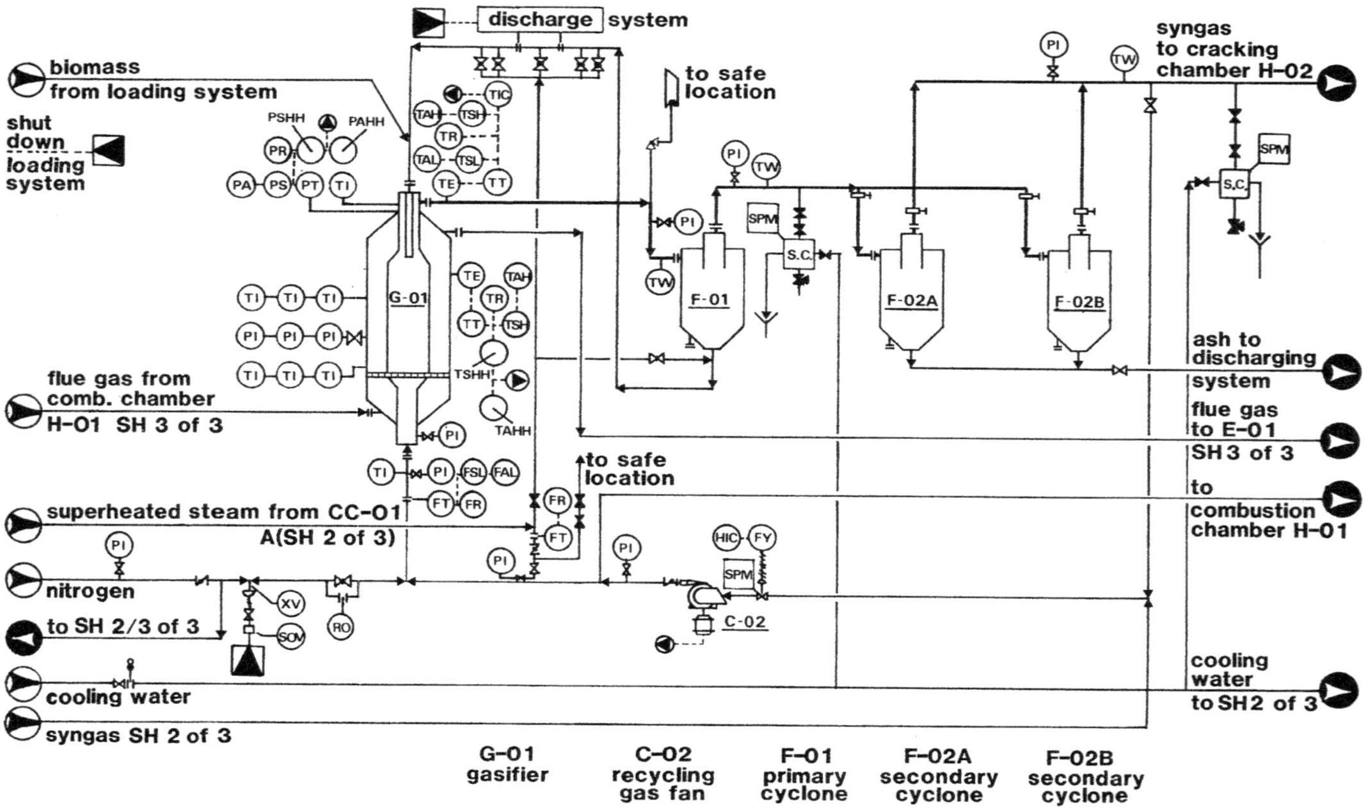

PIPING AND INSTRUMENT DIAGRAM - GASIFIER AND CYCLONES

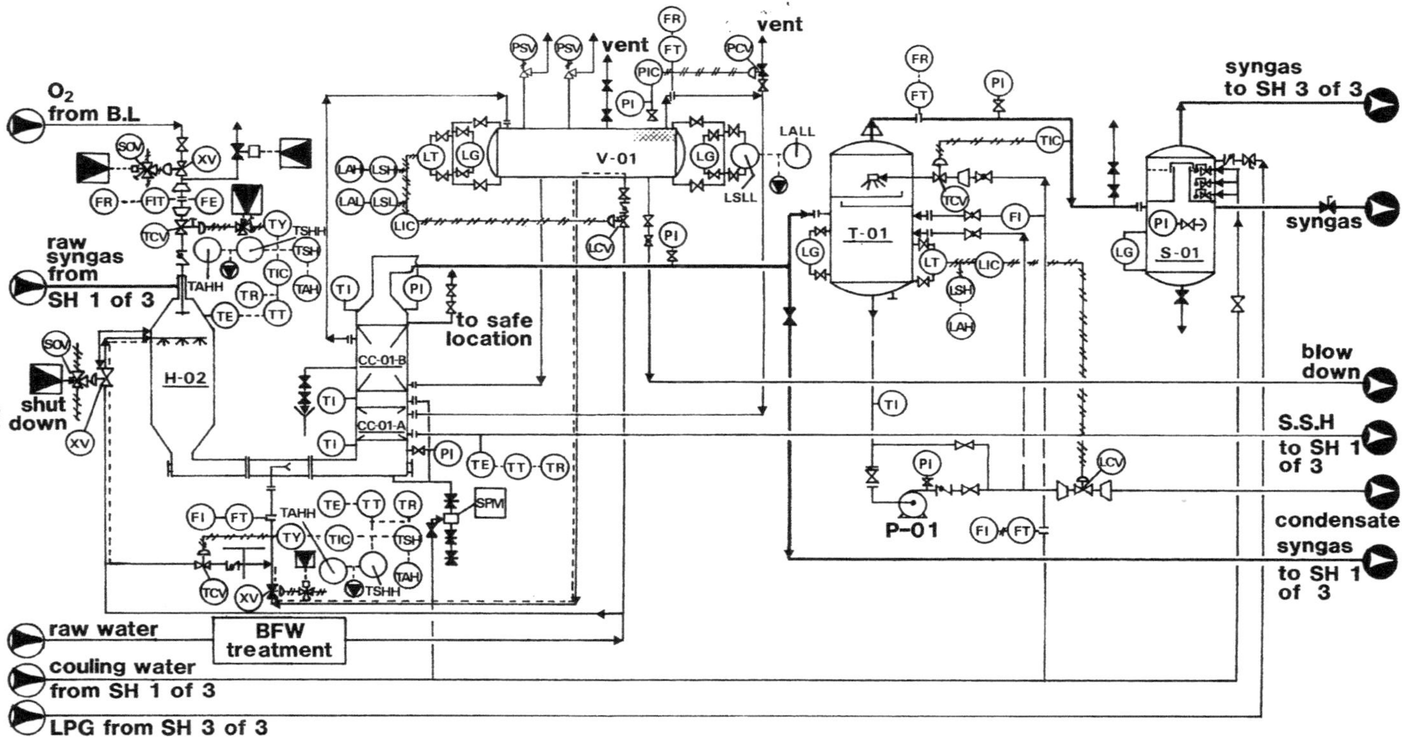

H-02 cracking chamber | Q-01 quencher | CC-01 A/B steam generation and S.S.H coil | V-01 steam drum | T-01 spray chamber | P-01 condensate pump | S-01 water seal

PIPING AND INSTRUMENT DIAGRAM - CRACKING CHAMBER - QUENCHER AND STREAM SYSTEM

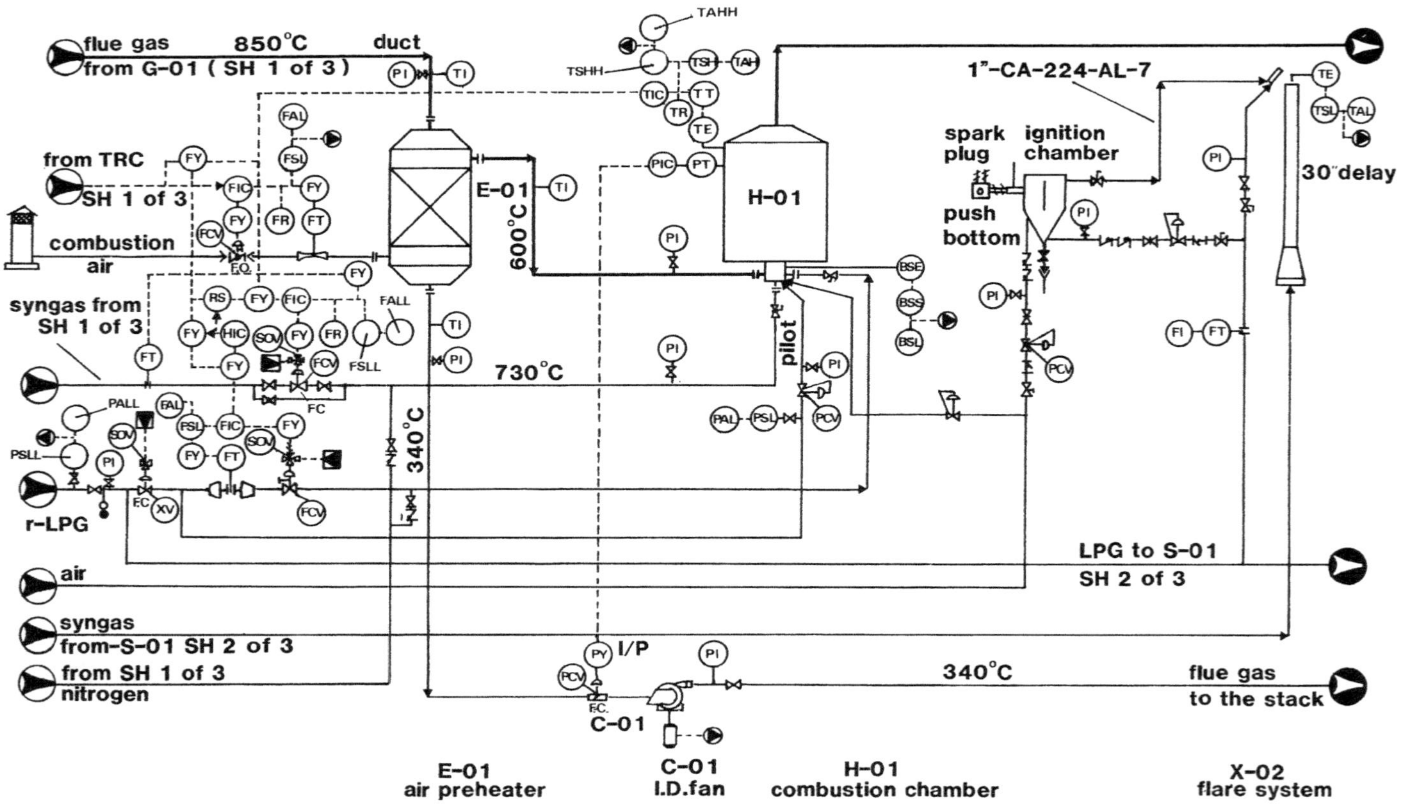

PIPING AND INSTRUMENT DIAGRAM - COMBUSTION CHAMBER - AIR PREHEATER AND FLARE SYSTEM

THE AVSA FLUID BED COMBUSTOR-GASIFIER PROJECT.

Author : H.A. MASSON

Contract number : ESE-R-087-B

Duration : 12 months

Total budget : FB 5.10^6 EEC participation : 40 %

Head of project : Dr. H.A. MASSON
Institute of Applied Mechanics
University of Brussels

Contractor : University of Brussels (Belgium)

Address : University of Brussels, C.P. 165
F.D. Roosevelt Avenue 50
B - 1050 Bruxelles - Belgium

Summary

A simplified design has been proposed for a small scale fluid bed combustor-gasifier system. Bench scale cold tests facilities have been used to optimise the operating parameters by a thermal tracing technique. Circulating rates as high as 1000 kg/sm^2 have been achieved. The selected fuel consists of biomass pellets. Their segregation tendencies have been determined in small scale units.
Steam gasification has been studied in a micropilot heated fluid bed. It produces a gas containing about 40 % H_2.
A pilot unit (200 kg/h) has been constructed and the combustion cell has been operated successfully. Combustion efficiencies as high as 97.8 % were obtained, while the thermal efficiency ranges around 83 %.

1. PROCESS DESCRIPTION

The AVSA process is part of the family of the double systems comprising a combustor and a gasifier.
In the combustor, solid, liquid or gaseous products are burned by air. It operates around 900°C. The heat generated by the combustion is stored as sensible heat of the solid. This one is sent back to the gasifier operating around 700°C. In this part of the process, sensible heat is partially used to satisfy the heat balance of the endothermal gasification process.
The system works in the fluidised regime to enhance the solid circulation between gasifier and combustor.
The principle of the AVSA process is illustrated at figure 1.
The expansion of a fluid bed increases with gas velocity. On the other hand, the pressure drop balances the weight of solid per unit section.
In a fluidised bed communicating vessel system, where the two branches are fluidised at different velocities, the height of the two branches would be different. If an overflow is allowed, the two levels would equilibrate, inducing a solid circulation.
This circulation principle has been applied in both of the parts.
The process presents thus four cells as shown in figure 2.
Two are affected to combustion, the two other ones to gasification.
The gasification and combustion parts are connected through immerged orifices, in such a way that a double communicating vessel system is realised.
In each part, the two cells are brought at different fluidisation velocities, to promote solid overflow from one cell to another. Solid goes by overflow from cell 1 to 2 ; from 2 to 3 through the immerged orifice ; from 3 to 4 by overflow ; and from 4 to 1 again through the immerged orifice.
Smooth operation of the system requires a great and regular transport of heat to the gasifier. The selected configuration realises heat transfer

1) through the sensible heat of the circulating solid ;
2) through the partition wall between combustor and gasifier ;
3) eventually by heat exchange pipes located in part 1 and 3, as defined on figure 2.

2. SOLID CIRCULATION TESTS ON COLD MODEL

2.1. Instrumentation

Gas flows are defined using Pitot tubes in each of the ducts leading air to the four cells of the system. Pressure drops through the distributor and the fluid beds are measured with water manometers. The localised pressure drops at the immerged orifice are also measured.
Solid circulation is studied by injecting a pulse of hot solid (200 g ; 150°C) at the top of cell 4. The thermal wave is detected in the several parts of the unit by small thermistors. The solid dispersion mechanism is much more rapid than non stationary conductive heat transfer from grain to grain. The conduction process affects thus the thermistor responses only as a drift of the basis line.
Gas transfer between the combustor and the gasifier has been investigated by injecting continuously CO_2 in the combustor part and by detecting the CO_2 concentration above the combustion and gasification parts of the unit. The CO_2 analyser used is a non dispersive I.R. analyser (Hartman and Brown).

2.2. Preliminary tests

Closing the communication holes, we have analysed the solid circulation in

one cell as function of gas velocity. As shown on figure 3, the gas velocity will at least equal 1.5 U_{mf} to achieve dispersion of the thermal wave through all the cells in a reasonable time interval.
In a second step, the immerged communication holes have been opened and cycling experiments have been conducted.
The cycling of the solid is proved by successive detection of the thermal wave in the four cells before coming back to the first.
A cycling time is defined as the time interval between two maxima detected in the injection cell.
Some significant results are shown on figures 4 and 5. The results clearly show that the gas velocity in the low velocity cells must exceed 1.5 U_{mf}. On the other hand, the expansion of the layer, which is the motor of the process, increases much more sharply near U_{mf} than at higher velocities ; further in this work, 1.5 U_{mf} has thus been selected as the velocity in the low velocity cells.

2.3. Effect of operating parameters (defined on figure 6)

2.3.1. Solid transfer mass flow (fig. 7 and 8)

The specific mass flow is defined as :

$$m = \frac{\text{(mass of solid in the system)}}{\text{(cycling time) (communicating holes area)}}$$

m fluctuates between 200 and 1000 kg/sm^2 and presents a maximum for H_0/H = 0.92 ; m increases with the gas velocity in the high velocity cells and as a first approximation does not depend on the shape of the communication orifices but only on its surface, or reduced surface s/S as defined on figure 6.

2.3.2. Pressure drop through the communication orifices

At figures 9 and 10, a correlation is presented between the pressure drop at the communication orifice and the solid mass flow rate.
A square root law describes quite well the results. This fact allows to measure, after calibration, the solid transfer rate by a conventional differential pressure sensor.

2.3.3. Gas transfer between combustor and gasifier (fig. 11 and 12)

Between 20 and 40 % of the gas flow is exchanged between the two parts of the system. Further work is needed on this point to reduce the gas exchange rate to a more reasonable level.

3. THE FUEL USED

Biomass pellets have been selected. Their main characteristics are summarised in table I.
The handling of large pieces of fuel in a fluid bed of fines leads to eventual segregation (or demixing) problems.
This aspect has been investigated using a radioactive marking technique.
The cumulated presence time of a Co^{60} grain glued on one piece of fuel has been detected using a NaI cristal associated to a photomultiplier and dedicated electronics as shown on figure 13.
The ratio of the cumulated presence time over the total operating time is an estimator of the presence probability of the grain at the measurement level.
The presence probability profile as function of height reveals directly any floating or sinking tendancy of the fuel grain.
Figure 14 shows results obtained when a biomass pellets is put in a fluid

bed of sand, porous clay or coal tailing ashes.
This last material leads to the most uniform presence probability profile and has been selected through this study.

4. FLUID BED VAPOGASIFICATION OF BIOMASS PELLETS

These tests were conducted at 700°C in a 25 cm I.D. electrically heated fluid bed. The gas stream used is a mixture of steam and nitrogen. Gasification products are analysed by a chromatographic method.
Table II gives typical gas compositions as observed, while table III is a summary of experimental conditions and mass balance of the process.
While for H_2 and O_2 the mass balances close quite perfectly, the C balance is satisfied at only 65 %. This has been explained under more by : carbon accumulation in the bed ; tar formation, whose weight has not been included in the mass balance ; inaccuracy in fuel elementary analysis. Finally the calculations are done, admitting equilibrium of the water shift reaction. This statement is probably an oversimplification when operating at a temperature as low as 700°C.

5. THE PILOT FACILITY AND THE COMBUSTION TESTS

5.1. The facility

A schematic view of the pilot unit is given at figure 15.
It consists of two parts, each having a square section of 1 sqm. One is dedicated to combustion ; the other to gasification.
During the present program duration, only the combustion cell has been operated.
Feeding is done by an overbed gravity system. Ash discharge is done through a pneumatic system. Air is injected through sparge tubes. Start up is achieved with natural gas premixed with air in the sparge tubes and lighted up using a pilot flame above the bed. The bed is equipped with an immerged and a convective heat exchanger.
Details of the instrumentation used for evaluation of the heat and mass balances are given in figure 16.
The combustion cell is equipped with control loops for bed level, temperature, air excess and steam temperature.
Alarms are generated by an on line microcomputer when bin level, air excess, bed temperature and level, steam temperature or pressure come out of range.

5.2. Typical results

As expected from radioactive cold tests, no segregation has been observed when burning biomass pellets in a bed of coal tailing ashes.
Figures 17 and 18 give the Sankey diagrams for heat and mass balance in the combustion cell.
Combustion efficiency range around 97.8 % while a thermal efficiency as high as 83 % has been obtained.

6. CONCLUSIONS

1) A new combustor gasifier system has been investigated. It is characterised by its compacity, simplicity and reliability.
2) Cold tests models have shown that mass transfer rates as high as 1000 kg/sm^2 are possible between the combustor and the gasifier.

The main operating parameters have been identified and optimised. It has been shown that the solid transfer rate is measurable by a simple differential pressure measurement, which is very attractive for inclusion in a regulation loop.

3) However 20 to 40 % of the gas is exchanged between the combustor and the gasifier. Further work is needed to reduce this gas exchange.

4) Segregation of large pieces of fuel in a fluid bed is best studied by a radioactive marking technique. An acceptable solution is to adapt the bed density to achieve good mixing of the fuel elements.
In the case of biomass pellets, coal tailing ash is a good fluidising medium.

5) A pilot plant, able to handle 250 kg/h of biomass has been constructed. At the present time, only the combustion cell has been operated, with a combustion efficiency of 97.8 %.

Size - shape : cylinder (ϕ = 1 cm ; L = 2-5 cm)	
Real density : 1200 kg/m^3	
% water (mass) : 4.9 %	Elementary composition
% ash (mass) : 0.3 %	C = 49 %
% volatiles (mass) : 85.3 %	H = 6 % O = 44 %
% fixed C (mass) : 9.5 %	N = 1 %
HHV : 4282 Kcal/kg	
Origin : VALBOIS, to SAINT-HUBERT (Belgium)	

Table I. Characteristics of biomass pellets.

H_2	37.5
N_2	18.5
CH_4	7.4
CO	9.4
CO_2	25.6
C_2H_4	1.8
C_2H_6	0.2
HHV :	2504.5 Kcal/Nm^3

Table II. Gas composition (% volume).

REACTION PARAMETERS

Flowrate air (Nm^3/HR ; 0°C, 1 atm) =	0.00
Air pressure (bar rel) =	0.00
Flowrate oxygen (Nm^3/HR) =	0.00
Oxygen pressure (bar rel) =	0.00
Flowrate steam (kg/HR) =	3.55
Flowrate solids (kg/HR) =	1.50
Flowrate nitrogen (L/HR) =	280.63
Ambient temperature (°C) =	20.00
Flowrate condensate (kg/HR) =	3.60
Flowrate fly ash (kg/HR) =	0.05
Flowrate char (kg/HR) =	0.00
Ambient pressure (bar) =	1.00

		C	H	O	N	Ash
	kg/HR	kg/HR	kg/HR	kg/HR	kg/HR	kg/HR
Air	0.000	0.000	0.000	0.000	0.000	0.000
Oxygen	0.000	0.000	0.000	0.000	0.000	0.000
Nitrogen	0.351	0.000	0.000	0.000	0.351	0.000
Steam	3.550	0.000	0.394	3.156	0.000	0.000
Solids waf	1.345	0.659	0.081	0.592	0.013	0.000
Moisture	0.135	0.000	0.015	0.120	0.000	0.000
Ash	0.020	0.000	0.000	0.000	0.000	0.020
TOTAL	5.401	0.659	0.490	3.868	0.365	0.020
Total gas	1.529	0.000	0.000	0.000	0.000	0.000
H_2	0.051	0.000	0.051	0.000	0.000	0.000
O_2	0.000	0.000	0.000	0.000	0.000	0.000
N_2	0.365	0.000	0.000	0.000	0.365	0.000
CH_4	0.086	0.064	0.021	0.000	0.000	0.000
CO	0.198	0.085	0.000	0.113	0.000	0.000
CO_2	0.781	0.213	0.000	0.568	0.000	0.000
C_2H_4	0.044	0.037	0.006	0.000	0.000	0.000
C_2H_6	0.005	0.004	0.001	0.000	0.000	0.000
Condensate	3.600	0.000	0.400	3.200	0.000	0.000
Fly ash	0.045	0.027	0.000	0.000	0.000	0.018
Char	0.000	0.000	0.000	0.000	0.000	0.000
TOTAL	5.174	0.431	0.479	3.882	0.365	0.018
BALANCE	95.805	65.485	97.823	100.361	100.000	87.794

Table III. Vapogaseification mass balance.

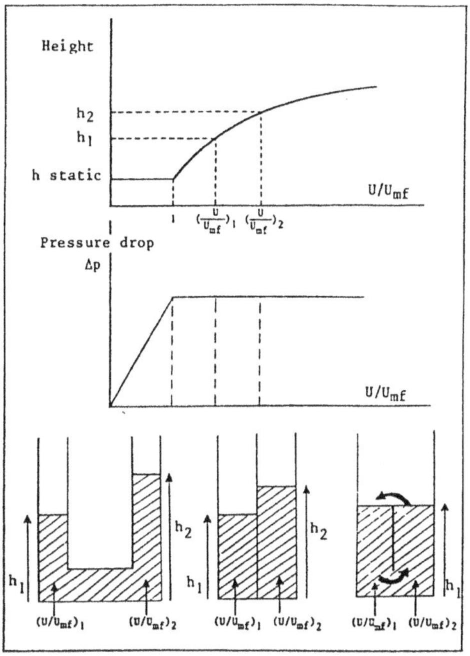

Figure 1.

Principle of the solid circulation.

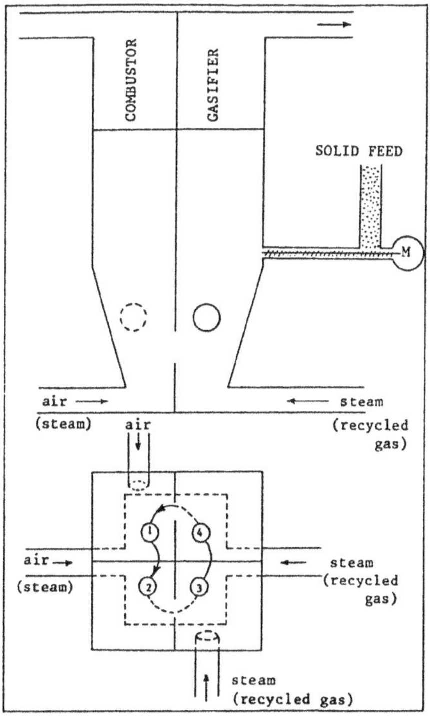

Figure 2.

Schematic view of the AVSA process.

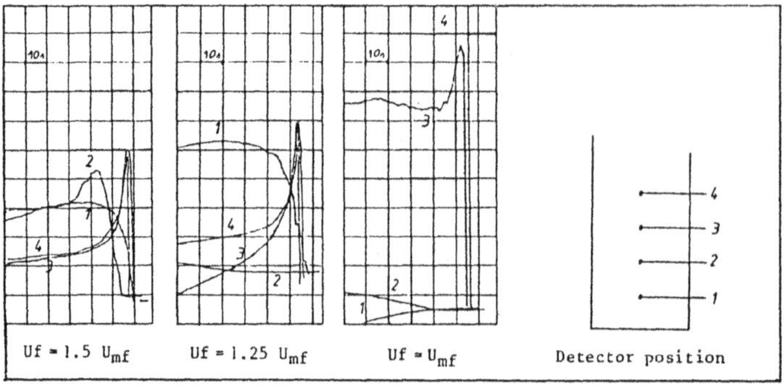

Figure 3.

Thermal pulse in one cell.

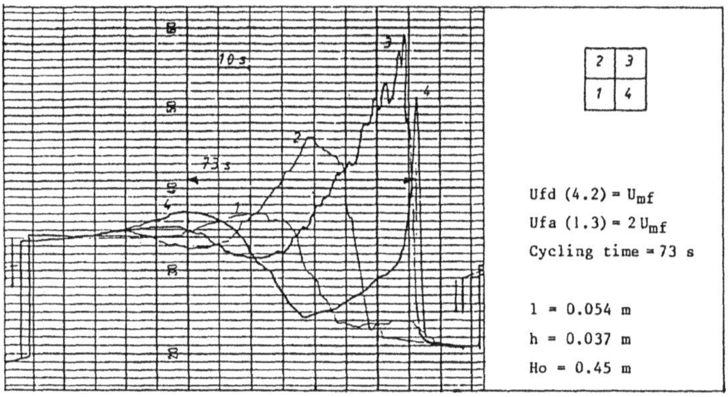

Figure 4.

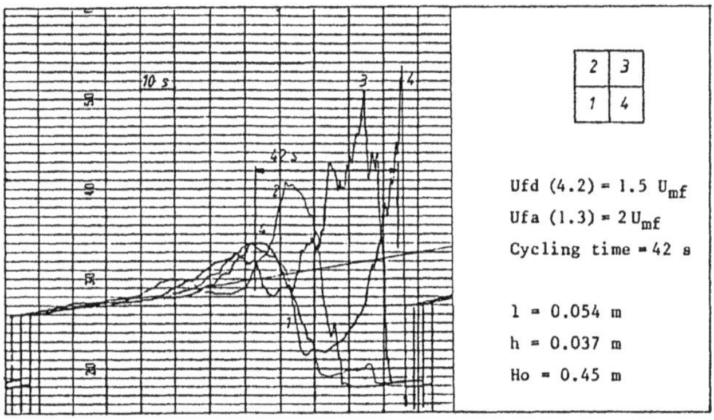

Figure 5.

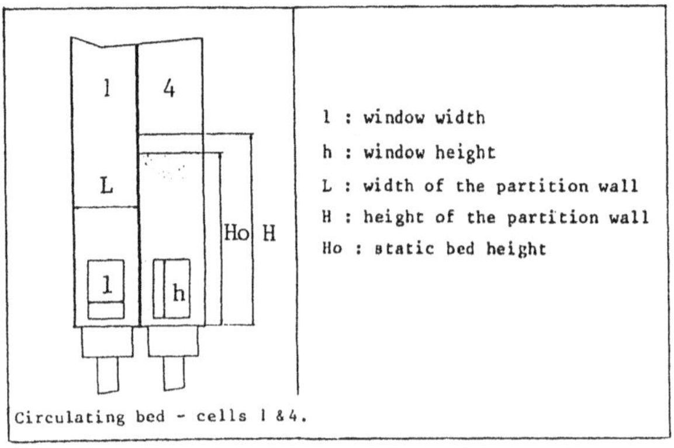

Figure 6.

Identification of the geometrical parameters.

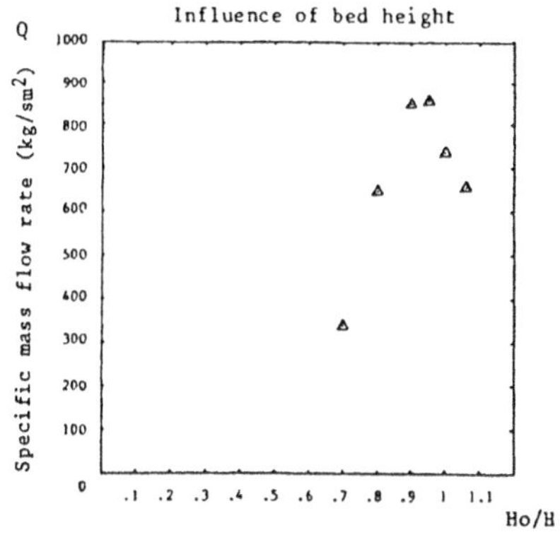

Bed height / partition wall height

△ S = l × h = 5.4 × 2.5 = 13.5 cm^2

Uf/U_{mf} = 2.5 ; H = .5 m

Figure 7.

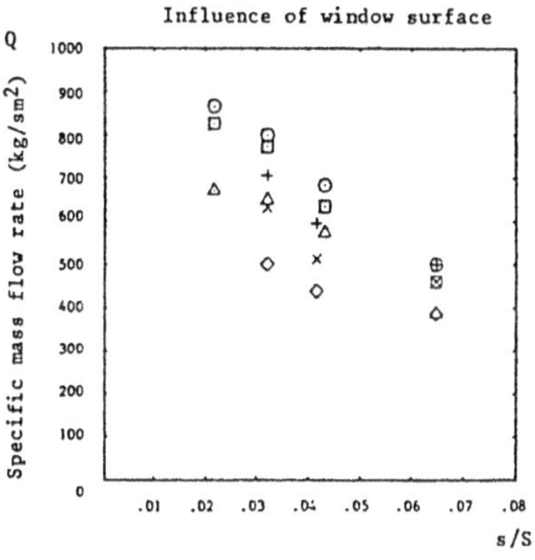

Window area / wall area

△ Uf/U_{mf} = 2 (l=.054 m)	◇ Uf/U_{mf} = 2 (h=.075 m)
⊡ Uf/U_{mf} = 2.5 (l=.054 m)	× Uf/U_{mf} = 2.5 (h=.075 m)
⊙ Uf/U_{mf} = 3 (l=.054 m)	+ Uf/U_{mf} = 3 (h=.075 m)

S : partition wall area = .0625 m^2
L : partition wall width = .125 m
H : partition wall height = .50 m
Ho : static bed height = .45 m

s : window area
l : window width
h : window height
Ho/H = .9
U_{mf} : minimum fluidisation velocity = .22 m/s

Figure 8.

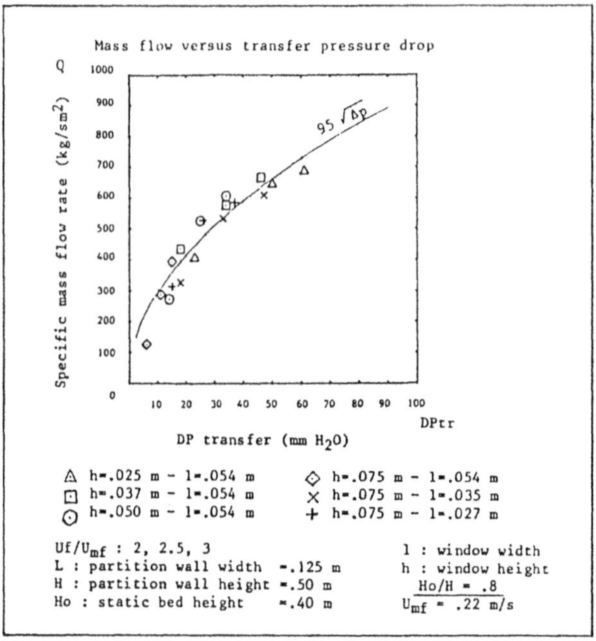

Figure 9.

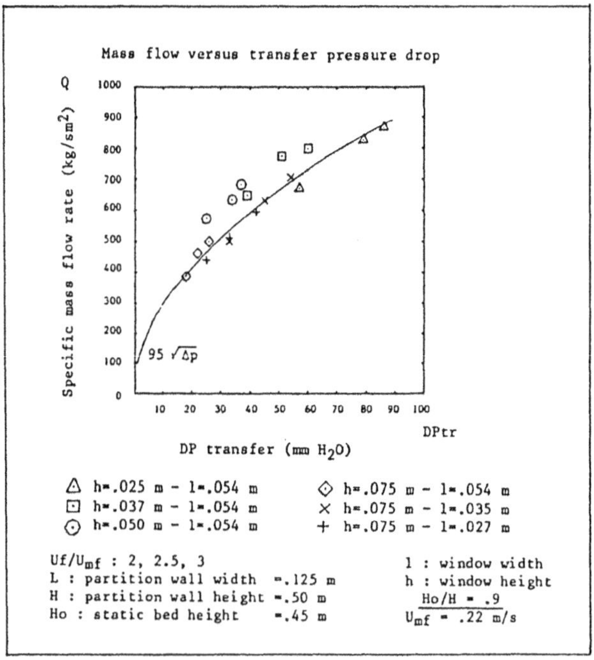

Figure 10.

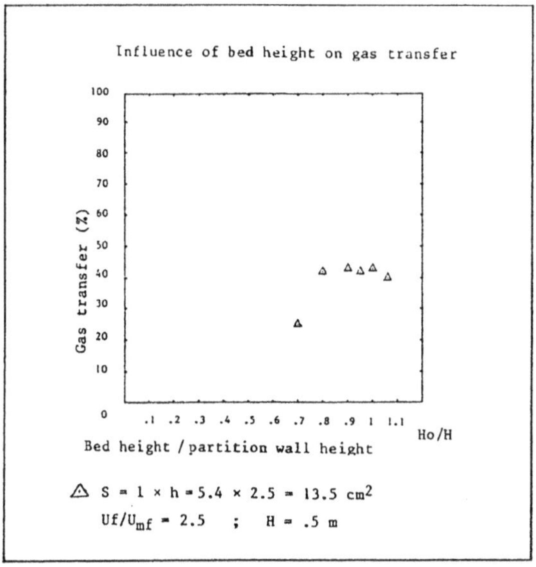

Figure 11.

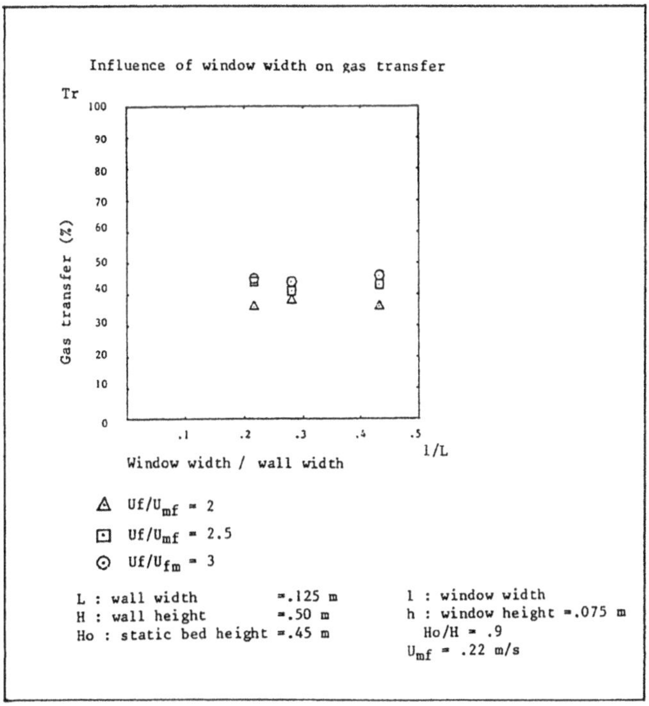

Figure 12.

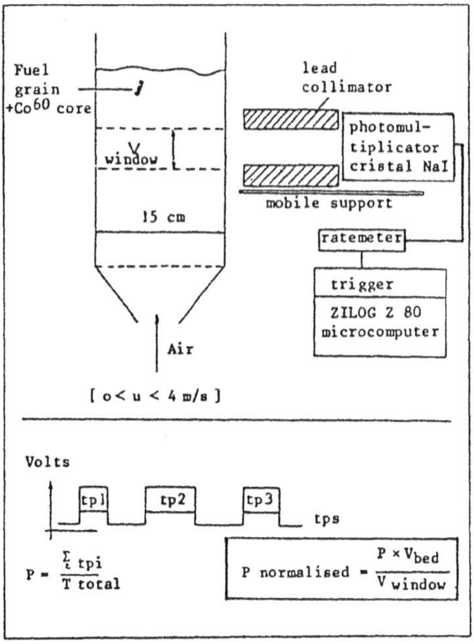

Figure 13. Presence probability (principle).

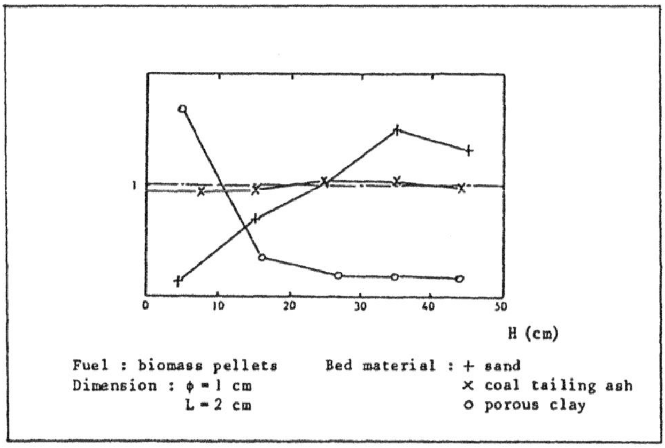

Figure 14. Presence probability of biomass pellets.

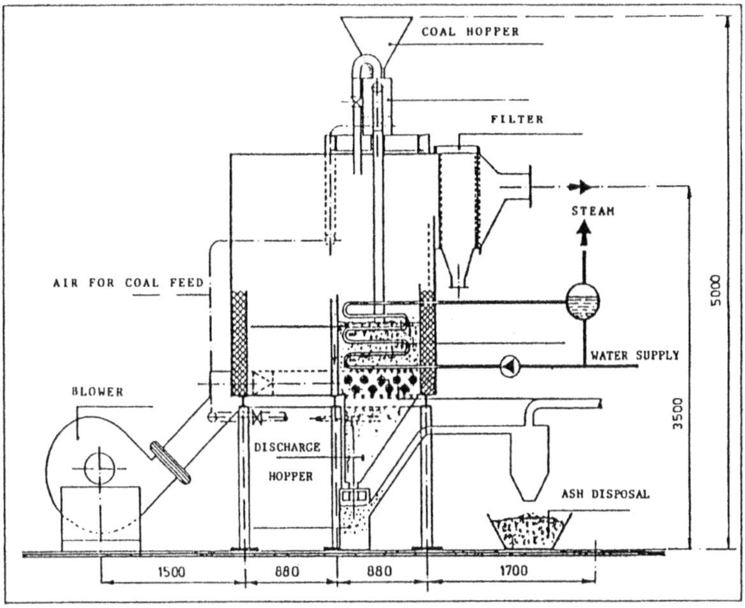

Figure 15. Experimental rig (2 × 1 MW).

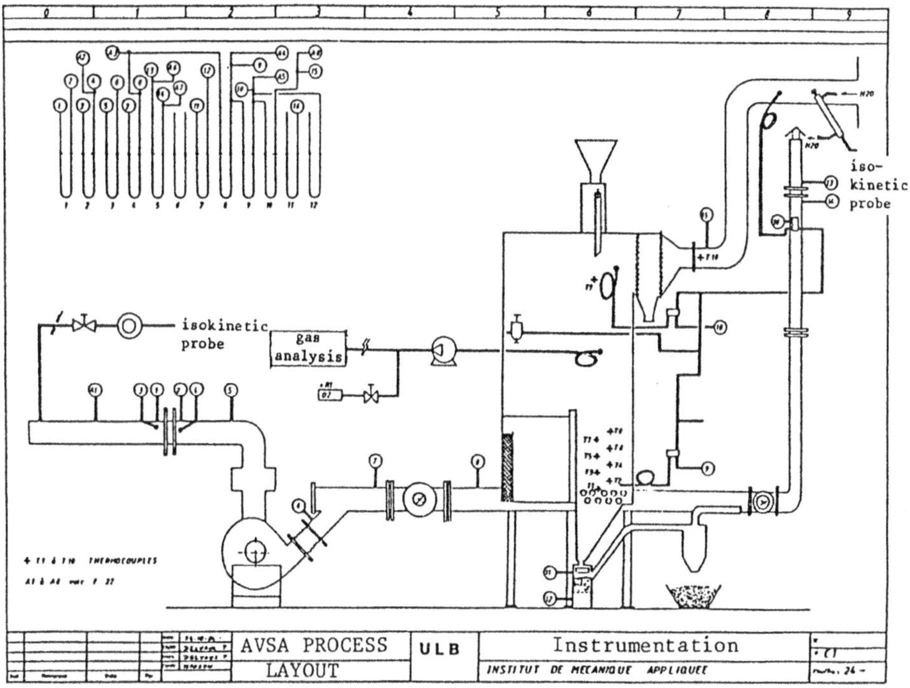

Figure 16.

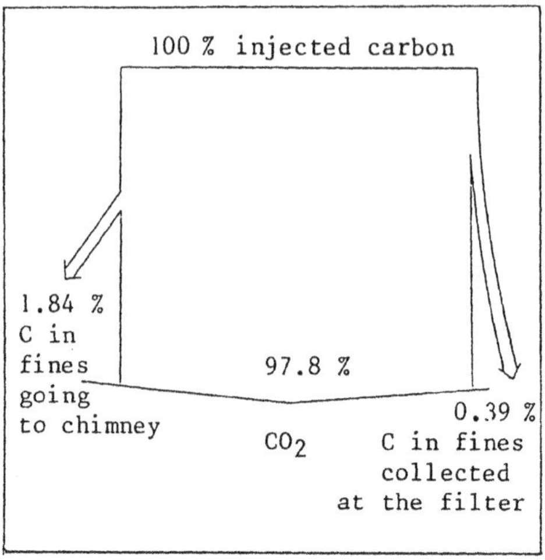

Figure 17. C Mass Balance (biomass pellets).

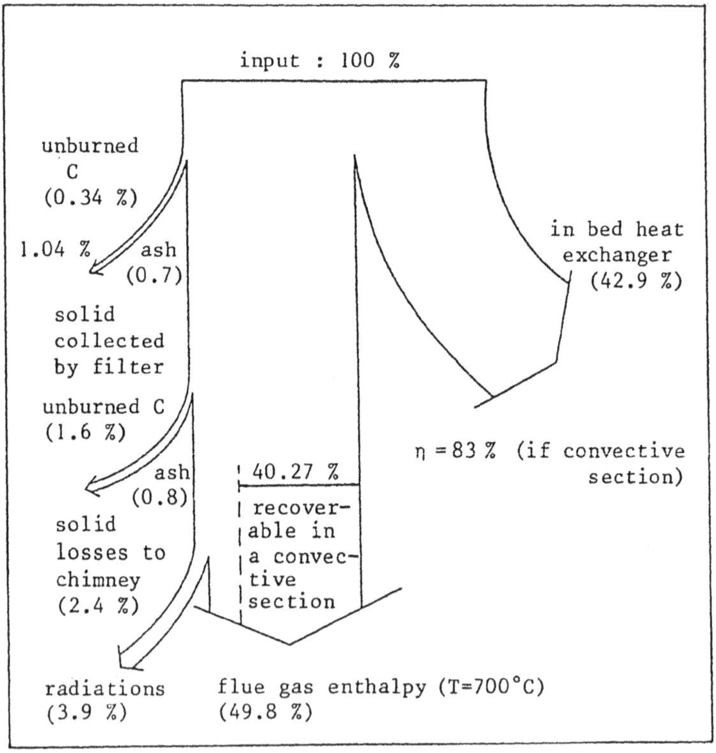

Figure 18. Heat Balance (biomass pellets).

HYDROGEN RECOVERY FROM GAS MIXTURES USING METAL HYDRIDES SUSPENDED IN SLURRY

Authors : K.J. PTASINSKI, A.A.C.M. BEENACKERS*, W.P.M. VAN SWAAIJ, R. HOLSTVOOGD

Contract number: ESE-R-062-NL

Duration : 24 months 1 January 1982 - 31 December 1983

Total budget : DFL 589.569 CEC contribution 50%

Head of project: Dr.ir. A.A.C.M. Beenackers

Contractor : Twente University of Technology

Address : Twente University of Technology
P.O. Box 217
7500 AE ENSCHEDE, The Netherlands

Summary

The recovery of hydrogen from producer gas obtained by air gasification is a key process for a viable "methanol from biomass" route. The present available technology is less feasible for the hydrogen recovery from producer gas and therefore a new method is proposed. This method is based on the absorption of hydrogen in a slurry containing finely dispersed particles of metal hydride in an inert liquid. The method has several attractive features such as easy continuous operation, improved heat transfer to metal particles and reduction of problems known to occur in discontinuously operated fixed beds, which is the conventional way of operation.

Under the present contract, a start has been made from just an idea of the new continuous hydrogen recovery process up to the pilot plant scale. Experimental results for a $LaNi_5$-silicone oil slurry system are presented. It is shown that the system is very reactive and the conversion rates per unit volume reactor can be at least as high as in dry systems with other conditions similar. The chemical absorption of hydrogen in a slurry appears to be controlled by the diffusion of hydrogen through the liquid film and the possibility of chemical enhancement of the mass transfer process is discussed. The continuous pilot plant designed under the contract and the experimental programme are described.

---

* Present address: Groningen University, Dept. of Chemical Engineering, Nijenborgh 16, 9747 AG Groningen, The Netherlands.

1. Current problems in methanol from biomass route

Various gasification schemes have been suggested for a viable "methanol from biomass" route. Oxygen and steam gasification are not yet considered to be feasible at the economically acceptable production level of 10^5 kg methanol/day. The former results in high air refrigeration costs [1] and the latter needs an extra secondary gasification or reforming step [2].

Air gasification could be the most simple method for small scale methanol production but hitherto existing schemes [3,4] result in high costs of nitrogen and carbon dioxide removal. To overcome these problems an alternative process is proposed, which is based on air gasification (e.g. Twente University of Technology moving bed gasifier [5,6]) and the successive recovery of carbon monoxide and hydrogen from the producer gas. Carbon monoxide can be selectively removed using a new process [7,8], which recently reached a commercial status.

Concerning the opportunities for economically feasible hydrogen recovery from lean gas mixtures, as producer gas, the picture is different. The existing pressure and temperature swing techniques are based on preferential absorption of all components except of hydrogen [9]. This makes these methods less attractive for air based producer gas, where hydrogen content is relatively low. Also selective diffusion of hydrogen through palladium membranes is less convenient because it requires a large pressure drop particularly at low partial hydrogen pressures [10].

A reversible absorption of hydrogen in hydridable metal alloys is a possible chemical alternative for selective hydrogen recovery. During the last fifteen years a great variety of metal alloys has been developed for hydrogen absorption [11-14]. The standard way of operation is the pressure swing technique in packed beds which is essentially a discontinuous process with several disadvantages as poor heat transfer, decrepitation of the particles, blocking of the beds, emission of fines and large forces on the reactor wall during hydrogen absorption.

Recently, we patented a new process [15] to overcome all these drawbacks by applying hydridable metal alloys finely dispersed in a liquid slurry, which is continuously recycled over an absorber and desorber column. This way, a continuous stationary process is obtained with good heat transfer, a low pressure drop and improved reactivity due to the possibility of using very small particles. An additional advantage of this process is the possibility to separate and compress hydrogen by low level heat and without use of a compressor. Typical ways of operation are any type of tray columns or trickle flow columns with a packing consisting of pipes for heat transfer.

Under the present EC contract a start has been made from just an idea of a new hydrogen recovery process up to the pilot plant scale. The experimental results obtained in a laboratory reactor for the slurry system: H_2-$LaNi_5$-PD5 silicone oil are presented in chapter 2, while in chapter 3 the continuous pilot plant is described.

2. Experimental results for the system: H_2-$LaNi_5$-PD5 silicone oil

The main object of this study is to demonstrate the possibility of a reversible absorption of hydrogen in a slurry containing finely dispersed metal hydride particles in an inert liquid. $LaNi_5$ hydride has been selected as it has a large absorption capacity and high absorption and desorption rates at relatively low temperature. Various partial ternary substitution can be made in $LaNi_5$ resulting in changes in plateau pressure thus greatly increasing its versatility. Al is especially potent in lowering the plateau pressure (Fig. 1) without reduction of the kinetics and the resistance to impurities.

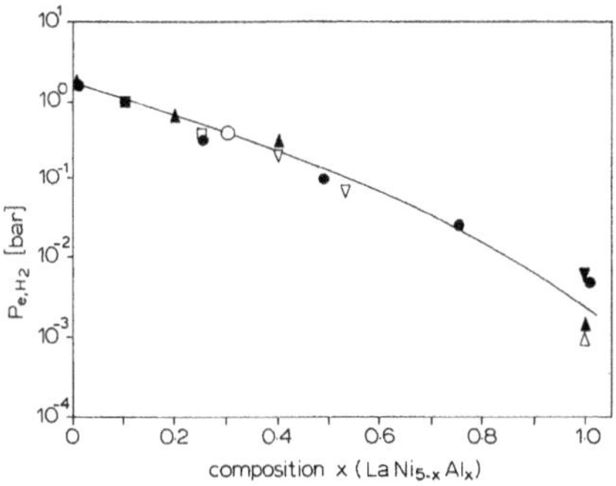

Fig. 1. The plateau pressure versus Al content for $LaNi_{5-x}Al_x$ at 250 °C (▲ -ref. 16, ▽ -ref. 17, ○ -ref. 18, △ -ref. 19, □ -ref. 20, ■ -ref. 21, ▼ -ref. 22, ● -ref. 23)

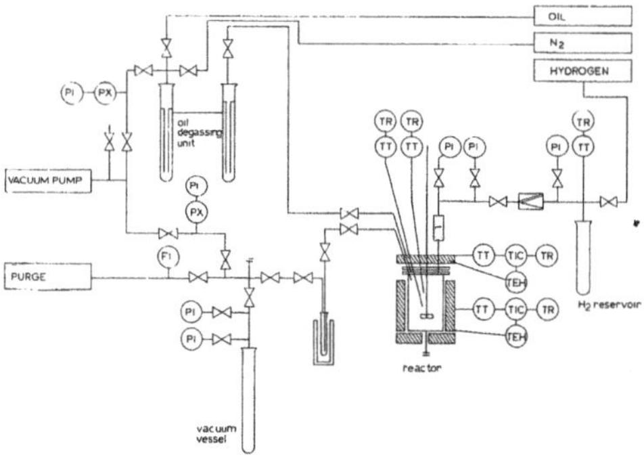

Fig. 2. Schematic diagram of the experimental apparatus

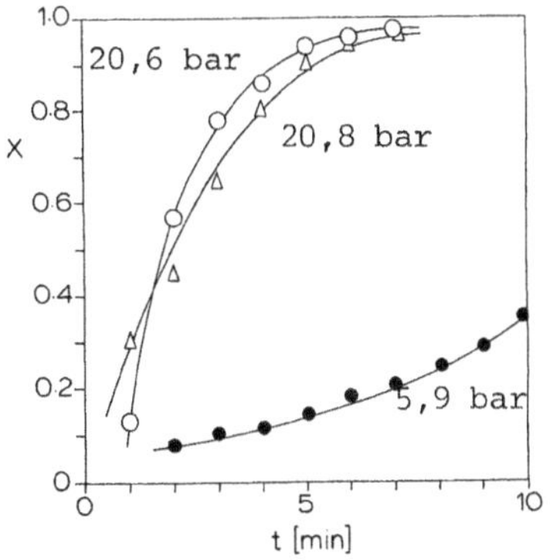

Fig. 3. Conversion of the dry $LaNi_5$ vs reaction time for various hydrogen pressures

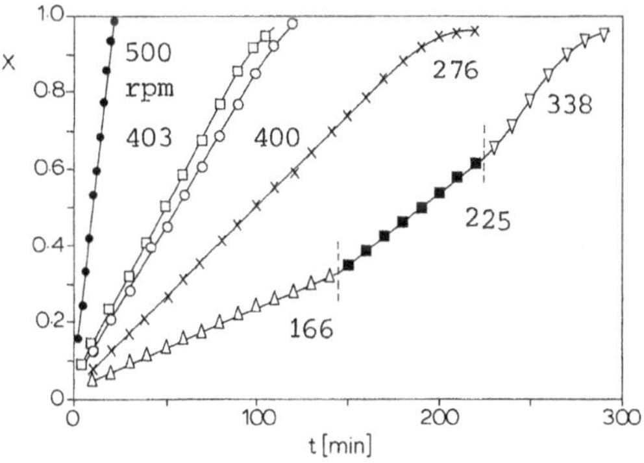

Fig. 4. $LaNi_5$ conversion vs absorption time for various stirrer speeds (T = 25 °C, p_{H_2} = 26 bar)

2.1. Apparatus and procedure

Figure 2 shows a schematic drawing of the experimental set-up. A batch stirred-cell reactor of 0.5 dm^3 volume was used for absorption-desorption experiments. Hydrogen was supplied from a pressure reservoir through a pressure regulator in order to keep constant pressure in the reactor. The amount of hydrogen absorbed in the metal hydride was calculated from the P-V-T readings of the reservoir and the reactor. After the absorption being completed, the hydrogen feed to the reactor was switched off and measurement of the desorption kinetics started. Hydrogen was removed from the reactor by reducing the reactor pressure and the volume of hydrogen removed was measured using a wet gas meter.

Ultrahigh purity hydrogen of $\geqslant$ 99.999% (Matheson) was used in all experiments. $LaNi_5$ was supplied by INCO Europe Ltd. Alloy was activated in the reactor following standard procedure [24]. The sample was also subjected to several hydriding-dehydriding cycles in order to measure pressure-composition isotherms and absorption rates. The absorption rates in dry metal powder (Fig. 3) cannot be discussed quantitatively because of the substantial temperature rise in the reactor. After the initial experiments with dry alloys,silicone oil PD-5 (Baysilone) was introduced into the reactor.

2.2. Kinetics of hydrogen absorption in slurries of $LaNi_5$

The absorption experiments were carried out within the following range of the parameters: temperature 25-70 °C, hydrogen pressure in the reactor 6-46 bar, stirrer speed 166-1200 rpm. In all experiments a slurry of 0.10 kg $LaNi_5$ and 0.26 kg PD-5 oil was used (28.4 wt per cent of metal).

The progress of the hydrogen absorption for different hydrogen pressures and temperatures was normalized using the relative degree of $LaNi_5$ conversion X, that is the ratio between the hydrogen amount absorbed and the maximum (equilibrium) amount at the given temperature and pressure.

Fig. 4 shows the typical relationship between the relative degree of $LaNi_5$ conversion X and the absorption time t. It can be seen that conversion reaches almost unity with the relationship X(t) being linear in the two phase metal-hydride region. The specific rate of absorption can be calculated from the slopes of the X(t) lines and expressed as kmoles of H_2 absorbed per unit of time (s), unit of hydrogen driving force (bar) and unit of surface area (m^2) of the liquid in the reactor (assuming the slurry surface to be flat). The assumption of the flat surface is only correct for a stirrer speeds $n \leqslant 400$ rpm.

It was shown that the absorption rate r depends on the stirrer speed (Fig. 5) and the temperature (Fig. 6) only, while it does not depend on the hydrogen driving force $\Delta p_{H_2} = p_{H_2} - p_{e,H_2}$. The absorption rates strongly increase with increasing mixing efficiency in the reactor, especially for $n > 400$ rpm. This can be explained by the experimentally observed rapid increase of the interfacial surface area with increasing rotation speeds above 400 rpm due to suction of gas bubbles into the slurry.

An important property of metal hydrides is the sensitivity to impurities. Fig. 7 shows the change of the final relative degree of $LaNi_5$conversion as a function of the number of absorption/desorption cycles (the dots for dry $LaNi_5$ are included too). After the first cycles where the absorption capacity decreases to about 0.95 of that for the fresh hydride, it further remains constant.

2.3. Kinetics of hydrogen desorption

Prospective applications of metal hydrides for hydrogen recovery require the hydride to be easily formed and decomposed. Therefore desorption experiments from the $LaNi_5$ slurry were carried out additionally. The desorption was examined within the range: temperature 25-66 °C, stirrer

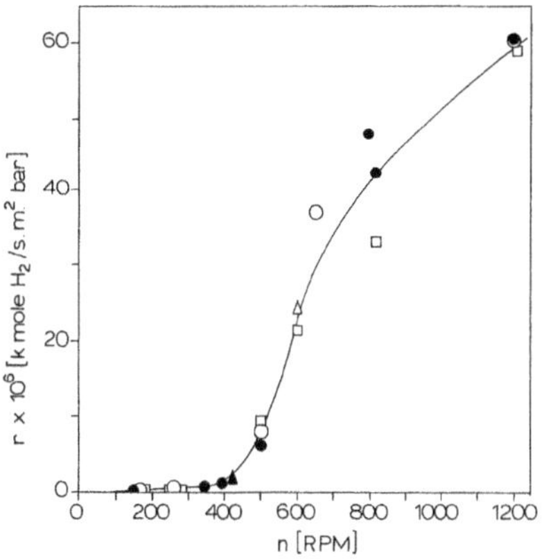

Fig. 5. Effect of stirrer speed on absorption rate (T = 25 °C, Δp_{H_2}: □ - 3.4, △ - 6.4, ○ - 13.4, ● -23.4, ▲ - 43.4 bar)

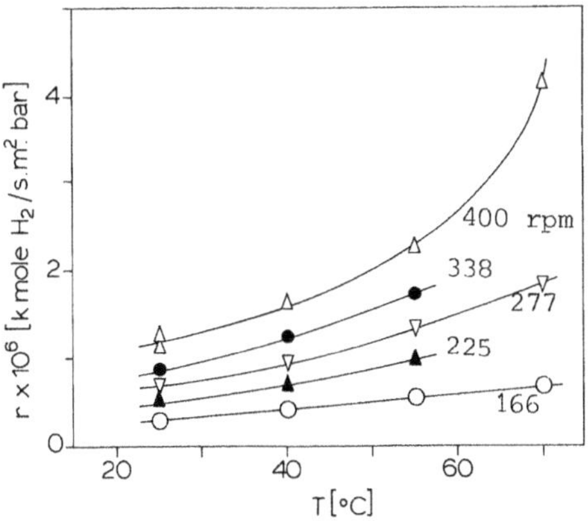

Fig. 6. Effect of temperature and stirrer speed on absorption rate

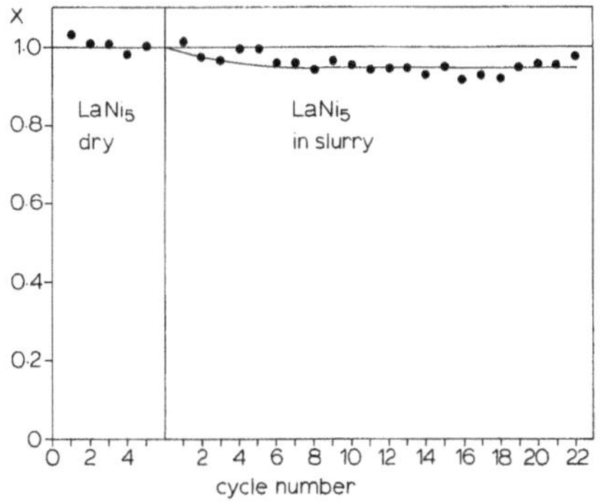

Fig. 7. Final conversion of $LaNi_5$ vs number of absorption cycles

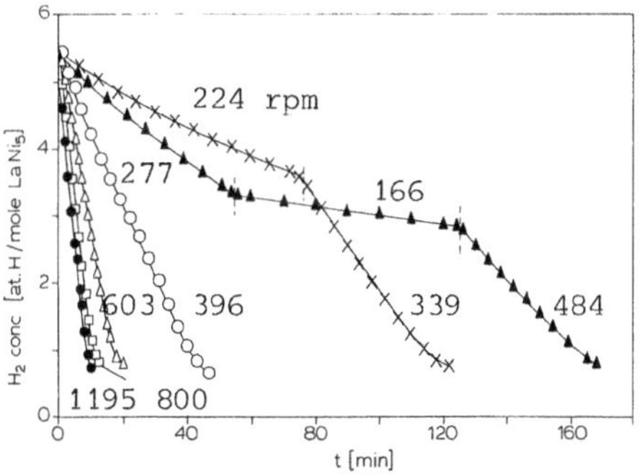

Fig. 8. Hydrogen concentration vs desorption time for various stirrer speed (T = 40 °C, p_{H_2} = 1.20 bar)

speed 166-1200 rpm and hydrogen pressure in reactor 1.08-1.87 bar.

Fig. 8 shows the typical variation of the hydride composition x as $LaNi_5H_x$ with desorption time. The desorption curves are almost rectalinear in the two phase region whereas the small deviation from the linearity can be explained by variation of the plateau pressure with hydrogen content. The initial and final sections of the desorption curves (for $x > 5.0$ and < 1.5 respectively) are outside the two-phase region and consequently deviate much more from linearity.

The effect of slurry temperature on desorption rates is very significant, as shown in Fig. 9. Desorption rates, r, in comparison with the corresponding absorption rates are much higher for the similar range of temperature and stirred speed. This is due to the fact that for desorption spontaneous bubbles cannot be avoided even not for rotation speeds < 400 rpm.

2.4. Discussion of results

The absorption of hydrogen in a slurry $LaNi_5$-oil PD5 in an example of gas absorption with reaction in slurries containing unsoluble fine reactive particles. Elementary steps of the process are: diffusion of H_2 in the gas film, diffusion of H_2 through the liquid film, diffusion of dissolved H_2 through the liquid-solid film and reaction between dissolved H_2 and metal hydride. Diffusion of H_2 through the liquid film and reaction can be parallel steps if very small particles are present in the liquid film.

To evaluate the enhancement the physical liquid-side mass transfer coefficients for the system: H_2-PD5 oil were measured in the same reactor and within the same range of temperatures, pressures and stirrer speeds as for the slurry. The comparison of k_L and k_L^o (Fig. 10,11) shows that the chemical absorption at higher temperatures is enhanced by the reaction between H_2 and metal hydride. This is analogous to common gas-liquid reactions, and has been observed for other slurry systems as well [25,26].

Physical mass transfer coefficients can be influenced by the presence of suspended solids. Experimental data on this topic are scarce and therefore we measured it by using inert methane as a model gas and otherwise the same system and experimental set-up as described above. Fig. 12 shows the comparison of mass transfer coefficients k_L^o for the clear liquid with those for the slurry, $k_{L,s}^o$ (the dotted lines). The $LaNi_5$ particles effect k_L^o considerably for the lower range of stirrer speeds $n < 250$ rpm, whereas for the higher range $n = 250$-400 rpm the influence of particles is much smaller. It was further found by visual observation of the mixing pattern in the reactor that an impeller speed of 250 rpm was minimally required to get just complete suspension, whereas the slurry was homogeneous for $n > 300$ rpm. Therefore at lower stirrer speeds the mixing efficiency in the liquid film is reduced in comparison to that for the clear liquid. For a homogeneous slurry, $n = 300$-400 rpm, the presence of particles lowered the k_L^o by a factor of about 0.9 only.

2.5. Conclusions

We have shown that hydrogen can be reversibly absorbed in a slurry of finely dispersed $LaNi_5$ in silicone oil PD-5.

The presence of the liquid causes a slight decrease in hydrogen capacity of about 5% per mol $LaNi_5$ in comparison with dry $LaNi_5$. The overall absorption rates in a slurry at a large effective gas-liquid interfacial area are of the same order of magnitude as the reaction rates with dry $LaNi_5$ in a fixed bed.

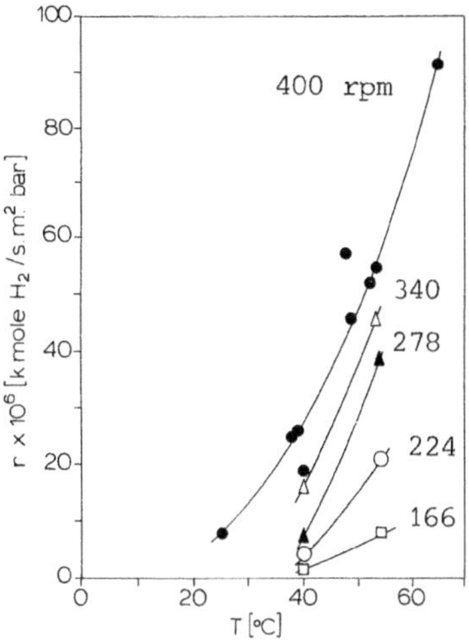

Fig. 9. Effect of temperature and stirrer speed on desorption rate, n < 400 rpm

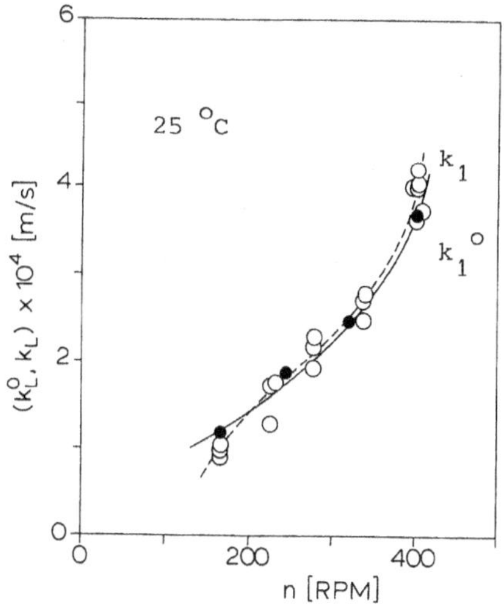

Fig. 10. k_1 in comparison with k_1^o at 25 °C

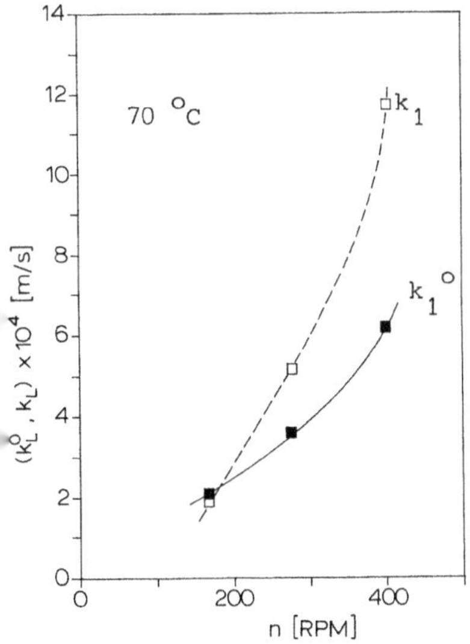

Fig. 11. k_1 in comparison with k_1^o at 70 °C

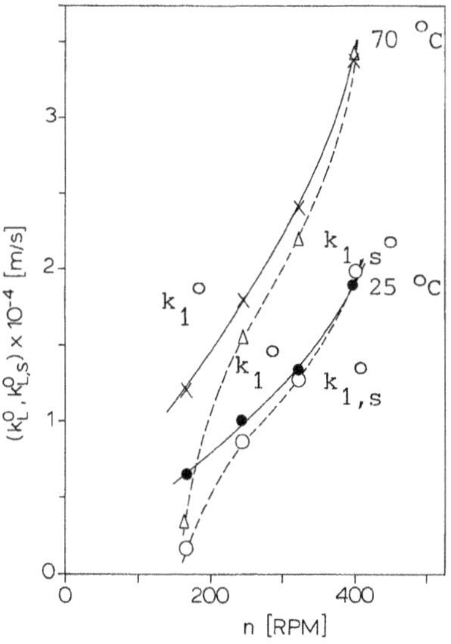

Fig. 12. $k_{1,s}^o$ (CH_4-$LaNi_5$-PD5 system) in comparison with k_1^o (CH_4-PD5 system) at 25 and 70 °C

The chemical absorption of hydrogen in a slurry seems to be controlled by diffusion of the hydrogen through the liquid film. For a homogeneous slurry in a stirred cell reactor the enhancement factor increases with increasing temperature whereas the physical mass transfer is slightly reduced by the presence of suspended $LaNi_5$ particles.

3. Continuous hydrogen recovery pilot plant

The selection of a reactor suitable for continuous hydrogen recovery with metal hydride slurries and the development of design rules for scaling-up of such a process are still fundamental problems in designing such a process. The knowledge on continuous gas-slurry reactors as published in open literature is rather limited and therefore we constructed a continuous pilot plant to investigate the hydriding behaviour of various metal hydride slurries in various contactors under the conditions close to industrial practice. These studies concern the determination of gas-liquid interfacial surface areas and mass transfer coefficients as well as a characterization of flow regimes and mixing properties for concentrated slurries under a wide variety of gas and liquid loads.

The pilot plant consists of an absorption and a desorption unit (Fig. 13). Different contactors as plate columns (up to three sieve, valve or bubble cap trays, respectively), bubble columns or packed columns with cooling elements can be tested within the absorber unit. The slurry (dehydrided in the stirred tank desorber) is pumped by a Leva membrane pump to the top of the absorber. The temperature in the absorber can be controlled either externally by oil or internally by cooling coils, while adiabatic operation is possible too.

In the absorber the slurry is countercurrently contacted with the gas mixture. To reduce the operating costs, the main part of the gas mixture is recycled to the absorber by means of a membrane compressor. The gas mixture is enriched in hydrogen just before the absorber. A system with a gas cooler and a heater is used in the gas loop to ensure a proper gas temperature.

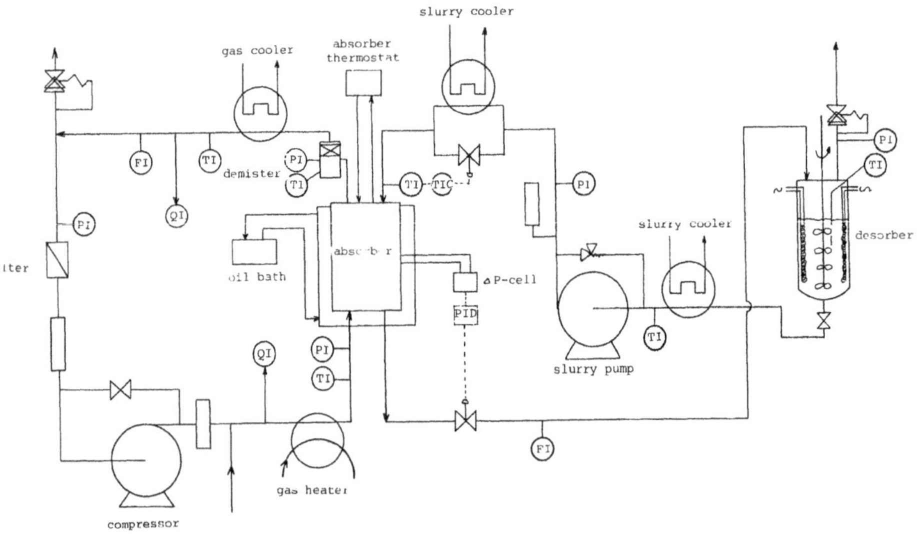

Fig. 13. Flowsheet of the pilot plant

Hydrogen concentration in the gas flow both before and after the absorber is measured by a gas chromatograph. The concentration of hydrogen in metal hydrides is measured by a specially designed analyser for hydrogen in slurries. The flow rates, temperatures and pressures of the streams are controlled and measured continuously around both the absorber and desorber. An additional small scale unit has been constructed for activation of fresh metal hydrides.

The design range of process conditions for the pilot plant is: temperature 20-300 °C, pressure 1-35 bar, slurry flow rate $0-50*10^{-6}$ m^3/s, gas flow rate 0-0.1 Nm^3/s, internal absorber diameter ⩽ 0.08 m and reactor height ⩽ 0.75 m.

4. Symbols

k_L	liquid side mass transfer coefficient for chemical absorption	m/s
k_L^o	liquid side mass transfer coefficient for physical absorption in the clear liquid	m/s
$k_{L,s}^o$	liquid side mass transfer coefficient for physical absorption in a slurry	m/s
n	stirrer speed	rpm
p_{H_2}	hydrogen pressure	bar
p_{e,H_2}	equilibrium hydrogen pressure	bar
R	specific absorption rate	kmole $H_2/s.m^2$
r	$R/(p_{H_2}-p_{e,H_2})$	kmole $H_2/s.m^2$ bar
t	time	min
T	temperature	°C or K
X	relative degree of $LaNi_5$ conversion	---
x	hydrogen concentration	at.H/mole hydride

5. References

[1] Reed T.B., A survey of biomass gasification, Vol. I: Synopsis and executive summary, SERI/TR-33-239-VI, Colorado, (1979).

[2] Kam A.Y. et. al. Preparation of a Cost Data Bank for DOE/Biomass Energy Systems Branch, DOE/ET/20605-4, US Dept. of Energy, 1979.

[3] Hokanson A.E., e.o., Chemicals from wood, NTIS, PB-262 489 (1975).

[4] Rowell R.M., in: Progress in Biomass Converison, I. Acad. Press (1979) 117.

[5] Twente University of Technology, Final Report EEC-contract 662-78-IESN, Enschede (1981).

[6] Neth. Pat. Appl. 79 03893 (1979).

[7] Foo K.W. and Shortland I., Hydrocarbon. Proc., (May 1976), 149.

[8] Chem. Eng. (Dec. 1977), 122.

[9] Lancelin L. Rudelstrofer E. van Scholter M., Rev. Assoc, Fr. Tech. Pet. 238 (1976) j44.

[10] Cox K.E., Proc. 2nd World Hydrogen Energy Conference, Zürich (1978), avail NTIS, LA-UR-78-938; CONF. 780807-5.

[11] Hoffman K.C., Reilly J.J., Salzano F.J., Waide C.H., Wiswal R.H. and Winsche, W.E., Int. J. Hydrogen Energy, 1 (1976), 133.

[12] Reilly J.J., in: Cox K.E. & Williamson, K.D. (Eds.), Hydrogen, its Technology and Implications, CRC, Cleveland (1977), Chapter 2, vol. II.

[13] Wiswall, R., Topics Applied Physics 29 (1978) 201.

[14] Sandrock, G.D. and Snape, E., A.C.S. Symp. Ser. (1980), 116; (Hydrogen, Production and Marketing), 293.

[15] A.A.C.M. Beenackers and W.P.M. van Swaaij, Method for separation of a gas, e.g. hydrogen, from a mixture of gases, Neth. patent appl. No. 8201885 (1982).
[16] Mendelsohn M.H., Gruen D.M. and Dwight A.E., Nature (London) 1977, 269, 45.
[17] Mendelsohn M.H., Gruen D.M. and Dwight A.E., J. Less - Common Met. 1979, 63, 193.
[18] Huston E.L. and Sandrock G.D., J. Less-Common Met. 1980, 74, 435.
[19] Takeshita T., Malik S.K. and Wallace W.E., J. Solid State Chem. 1978, 23 271.
[20] Belkbir L., Gerard N., Percheron-Guegan A. and Achard J.C., Adv. Hydrogen Energy 1979, 1, 2589.
[21] Belkbir L., Joly E. and Gerard N., J. Less-Common Met. 1981, 81, 199
[22] Tauber A. and Finnegan R.D., Rare Earth Mod. Sci. Technol. 1982, 3, 493
[23] Percheron-Guegan A., Achard J.C., Sarradin J. and Bronoël G., Proceedings Int. Symp. on Hydrides for Energy Storage, Geilo, Norway, 1977., A.F. Andresen and A. Maeland, eds., pp 485 - 490. Pergamon Press, London 1978.
[24] Buschow K.H.J. and van Mal H.H., J. Less-Common Met. 1972, 29, 203.
[25] Alper E., Wichtendahl B. and W.-D. Deckwer, Chem. Engng. Sci. 1982, 37, 217.
[26] Pal S.K., Sharma M.M. and Juvekar V.A., Chem. Engng. Sci. 1982, 37, 327.

AUTHOR INDEX

9 789027 722126

GPSR Compliance

The European Union's (EU) General Product Safety Regulation (GPSR) is a set of rules that requires consumer products to be safe and our obligations to ensure this.

If you have any concerns about our products, you can contact us on ProductSafety@springernature.com

In case Publisher is established outside the EU, the EU authorized representative is:

Springer Nature Customer Service Center GmbH
Europaplatz 3
69115 Heidelberg, Germany

Batch number: 10809441

Printed by Printforce, the Netherlands